PERSPECTIVES ON
BIOINORGANIC CHEMISTRY

Volume 4 • 1999

TRIBUTE TO ROBERT W. HAY 1934–1999

by Dr. David T. Richens
(close colleague and friend)

I am most honored to have the opportunity to pay tribute at this very sad time to my colleague Bob Hay whom I have known and formed a close friendship with since 1983.

Robert Walker Hay was born in Stirling in 1934 and spent his early childhood there before his family moved to England. He attended Bolton Grammar School where, it should be noted, Professor Sir Harold Kroto was also a pupil at that time. A death in the family brought them back to Stirling where Bob finished his education at the High School. He then went to Glasgow University to study for his B.Sc. and Ph.D. degrees in chemistry, graduating Ph.D. in 1959 in carbohydrate chemistry.

Following a brief period in industry, Bob took up his first academic appointment in New Zealand at the Victoria University of Wellington where he taught both organic and inorganic chemistry. However it was for his work with Neil Curtis carrying out some of the earliest experiments on self-assembly reactions that he will be remembered. The Curtis–Hay tetraaza macrocyclic ligands, involving simple condensations of diamines with acetone, were some of the first such systems to be prepared. Indeed the preparation of one of these ligands still forms part of an undergraduate laboratory experiment in St. Andrews today, providing a facile route into the unique chemistry of macrocyclic metal complexes. During his 10 years in New Zealand Bob's family life really came to the fore with the birth of three of his four children.

In 1971, Bob returned with his family to the U.K. to take up an appointment as Reader in chemistry at the new University of Stirling where he was made a full Professor in 1986. In 1988 Bob, along with myself and Frank Riddell, transferred to St. Andrews to spend what turned out to be the last 10 years of his prolific and fruitful academic career during which he has given so much to the chemical community in terms of both science and service. Ironically, he was due to take his fully deserved retirement later this year. That being said, the idea of Bob not having some chemistry to write about even in his retirement would have seemed impossible.

Bob's research interests and knowledge across chemistry were great. Throughout his career he retained an interest in biomimetic chemistry, specifically the study of metal ion-promoted reactions and reactions of molecules activated by metal ion coordination. His early interests in carbohydrate chemistry inspired him to study metal ion catalysis of both peptide formation and hydrolysis as well as studies in inorganic reaction mechanisms. He was particularly interested in the mechanisms of base-catalyzed hydrolysis within metal complexes and the development of the so-called dissociative conjugate-base (DCB) mechanism for base-catalyzed substitution reactions at inert d^6 metal ions such as Co(III).

In more recent times his research encompassed valuable solution studies on the behavior of cisplatin and analogues relating to their action as anticancer agents. Ironically, cisplatin was one of the drugs Bob received as part of his chemotherapy treatment. He was also working on the development of the use of micelles and solid-supported reagent systems to enhance/immobilize metallo-catalysts, e.g. for the effective destruction of chemical weapons materials, an area in which interesting developments are being made.

My enduring memories of Bob are concerned with his prolific writing, his calm approach to problems, and the ease with which he could puncture pomposity. Despite working with a group of never more than 3 or 4 people he was principal author of more than 220 primary research papers, book chapters, and books, a testimony to his sharp intellect, measured approach, and realistic choice of research topics. "Keep things 'simple,'" he always told me. "Set yourself ambitious targets—yes, but with attainable goals at all stages." Fundamental to his philosophy was the hand-in-hand approach to teaching and research.

He found the time to be the coeditor of two different book series and author of two teaching texts, one shortly to be completed on "inorganic reaction mechanisms". His paperback *Bioinorganic Chemistry*, first published in 1984, was the first book to provide a "get started" approach to this vast topic. It has proved to be a best selling undergraduate course text worldwide having been published in more than a dozen languages including Russian and Japanese. In 1997, he helped launch, as scientific coeditor, the first and long awaited primary research journal devoted to the topic of inorganic reaction mechanisms.

In 1994 he was instrumental in setting up, under the auspices of the Royal Society of Chemistry, the first discussion group in Europe devoted to coordination chemistry. He also had a long association with the RSC inorganic reaction mechanisms discussion group having served as a secretary and chairman and having been present at every meeting of the group since 1972 until this year. Bob was instrumental in bringing the International Symposium on Macrocyclic Chemistry to St. Andrews in July 2000, a meeting which would have marked the start of his retirement.

We will all miss his friendship, his wise counsel, his calm and measured approach to teaching and research, and of course his vast experience in Departmental, Faculty, and University matters.

David Parker, friend and Professor of Organic Chemistry at Durham University, summed up Bob's character in a recent communication. "Bob was one of the very few: sincere and modest, yet intellectually sharp and innovative."

PERSPECTIVES ON BIOINORGANIC CHEMISTRY

Editors: ROBERT W. HAY
Department of Chemistry
University of St. Andrews

JON R. DILWORTH
Department of Chemistry
University of Essex

KEVIN B. NOLAN
Royal College of Surgeons in Ireland
Dublin

VOLUME 4 • 1999

JAI PRESS INC.
Stamford, Connecticut

JAI PRESS INC.
100 Prospect Street
Stamford, Connecticut 06901

ISBN: 0-7623-0352-2

Manufactured in the United States of America

CONTENTS

LIST OF CONTRIBUTORS

Jorma Arpalahti	Department of Chemistry University of Turku Turku, Finland
Philip J. Blower	Nuclear Medicine Department Kent and Canterbury Hospital Canterbury, England
Janice Bramham	Department of Biochemistry University of Edinburgh Edinburgh, Scotland
Graeme Card	Department of Crystallography Birkbeck College London, England
Tamás Gajda	Department of Inorganic and Analytical Chemistry Attila József University Szeged, Hungary
Yoshio Inouye	Department of Environmental Hygiene Toho University Chiba, Japan
Eiichi Kimura	Department of Medicinal Chemistry Hiroshima University Hiroshima, Japan
Tohru Koike	Department of Medicinal Chemistry Hiroshima University Hiroshima, Japan
Roland Krämer	Inorganic Chemistry Institute Chemisches Institute Westfalische Wilhelm University Munster, Germany

Peter F. Lindley	European Synchotron Radiation Facility Grenoble, France
Sushumna Prakash	Nuclear Medicine Department Kent and Canterbury Hospital Canterbury, England
Vjacheslav Zaitsev	Institute of Crystallography Russian Academy of Science Moscow, Russia
Irina Zaitseva	Institute of Crystallography Russian Academy of Science Moscow, Russia

INTRODUCTION TO THE SERIES: EDITORS' FOREWORD

The aim of this series is to provide authoritative reviews in the rapidly expanding area of bioinorganic chemistry. The series will present "state of the art" reviews covering the whole field of bioinorganic chemistry. As the subject is, by its very nature, interdiciplinary, the editors feel that there is a need for articles covering the many different aspects of the subject from medicinal chemistry to biophysical studies. Suggestions from readers regarding topics to be covered in subsequent volumes will be welcomed.

R.W.H.
J.R.D.
K.B.N.

PREFACE

The present volume is the fourth in the series and covers the topics: lithium in biology, the structure and function of ceruloplasmin, rhenium complexes in nuclear medicine, the anti-HIV activity of macrocyclic polyamines and their metal complexes, platinum anticancer drugs, and functional model complexes for dinuclear phosphoesterase enzymes. The production of this volume has been overshadowed by a very sad event—the passing away of the senior editor, Professor Robert W. Hay. It was he who conceived the idea of producing this series and who more than anyone else has been responsible for its continuation. A tribute by one of his many friends, Dr. David Richens, is included in this Volume.

I wish to express my gratitude to the authors of the chapters for their contributions, their patience during unavoidable delays, and their willingness to provide updates on their respective chapters when requested. I would also like to thank Fred Verhoeven, Production Editor, JAI Press, for his help and encouragement in the production of this Volume.

Kevin B. Nolan
Series Coeditor

LITHIUM IN BIOLOGY

J. Bramham

Perspectives on Bioinorganic Chemistry
Volume 4, pages 1–50.

ISBN: 0-7623-0352-2

1. INTRODUCTION

Lithium is a ubiquitous element in the environment: it occurs naturally in seawater at levels of the order of 10^{-7} g/g, it is relatively abundant in rocks and minerals at approximately 10^{-5} g/g, and it is naturally present in ultratrace amounts in plants and in animals, including humans, at less than 10^{-8} g/g [1]. A few desert plants can accumulate Li^+ at levels up to 5 mmol kg^{-1} dry weight [2]. The levels of lithium found in plants and animals is dependent upon the geographical location and appears to reflect levels in the local water supplies [3].

Lithium has no known essential biological function. Nevertheless, over a century ago, lithium salts and lithia spring waters were being administered for a diverse range of medical applications, including the treatment of gout and a plethora of other illnesses. In modern psychiatry, the serendipitous discovery by Cade of the benefit of the lithium cation in the treatment of manic patients is now almost legendary [4]. The controversy surrounding lithium as a therapeutic agent, which arose primarily from its unfortunate toxicity [5], was responsible for the initial restrictions to its use and, therefore, the necessary research into its physiological actions. A resurgence of interest occurred around 15 years later with the work of Schou and others, and the efficacy of the lithium ion as a prophylactic against recurrent bipolar affective disorder, better known as manic depression, and in the treatment of acute mania is now widely appreciated [6,7]. Increased knowledge and awareness of the toxic manifestations of this cation have led to specific treatment regimes, involving dosage and patient monitoring, to minimize the chance of such toxicity developing, while gaining maximum benefit from the drug.

In industrialized countries it is estimated that around one person per one thousand is currently using lithium successfully for the prevention of relapses of manic-depressive episodes. To a lesser, but increasing extent, lithium salts are prescribed, sometimes controversially, for a number of other neurological disorders including unipolar depression, schizophrenia, childhood behavioral problems, aggression, and chronic alcoholism. The claims that lithium can also act as an antisuicide drug have been reviewed recently [8]; it is thought that its efficacy in this case is probably due to its prophylactic effect in affective disorders, although it may also be related to its antiaggressive properties. Additionally, lithium is employed against various somatic illnesses such as herpes simplex virus, leukocytopenias, inflammatory diseases, and dermatoses.

Although numerous diverse theories have been proposed, the fundamental biological role of the lithium cation, Li^+, and its therapeutic mechanism of action in relieving bipolar disorder are still not established. There is no direct evidence for any essential function for lithium in the human body and no unique or specific bonding for the cation with any biological molecule has been found. Those studies reporting effects upon receptor binding employ concentrations of Li^+ significantly above any therapeutically relevant levels and the effects are probably due to displacement of Na^+ or other cations. Despite this lack of specific interaction, in all organisms treated with Li^+, from viruses to plants and animals, the distribution of the cation is widespread and there is a diverse abundance of physiological effects. This is hardly surprising when one considers the biological distribution and physiological behavior of the other, chemically similar metal cations that are essential to sustain life. The larger alkali metal cations, Na^+ and K^+, and the alkaline earth metal cations, Mg^{2+} and Ca^{2+}, are generally present in much higher concentrations in organisms than those found for Li^+ at therapeutically relevant levels. Like those of Na^+, K^+, Mg^{2+}, and Ca^{2+}, the biochemical interactions of Li^+ are predominantly ionic, and many of the observed effects are a consequence of Li^+ mimicking, or substituting for, these essential cations.

Research to determine the therapeutic mechanism of action of Li^+ continually uncovers more physiological systems which are affected by this ion, resulting in ever expanding fields of research, explanations for specific toxic manifestations or side effects, and serendipitous therapeutic uses. Li^+ has been shown to influence numerous components within the central nervous system, each of which, or indeed any combination of which, may be the target for the therapeutic efficacy of this cation in psychiatric disorders. Several key aspects of intracellular signaling pathways, which are indirectly involved in the regulation of neurotransmitter function, are affected by the presence of Li^+, including adenylate cyclase activity, phosphoinositide metabolism, and guanine–nucleotide binding proteins. Many studies also show Li^+-induced effects upon the levels of neurotransmitters and hormones, and/or their metabolites and precursors. In addition, viral replication, immunology, dermatology, and leucocyte metabolism are all affected to some extent by this metal cation. The effect of Li^+ upon blood cell production has led to much exciting interest in its potential in bone marrow transplants and in the treatment of autoimmune disease. Some of these observed stimulatory and inhibitory effects of Li^+ have also led to its use as a powerful tool for studying biochemical pathways.

This cascade in research is reflected in the increasing amount of literature published each year on the biochemical behavior of the lithium ion and the aim of this article is to bring together much of this information, emphasizing the ubiquity of this small, apparently nonessential, metal cation in biology.

2. CHEMISTRY AND BIOCHEMISTRY

Lithium is the lightest solid element and is the least reactive of the alkali metals. The isotopic composition of natural lithium is 92.58% ^7Li and 7.42% ^6Li. As with the other alkali metals, the chemistry of lithium is predominantly that of the monovalent cation. Practical uses for lithium have been found in fuel cell technology due to its relatively high electrode potential (lithium batteries), and it performs an important role in modern organic synthesis, including the use of organolithium reagents as anionic polymerization catalysts. Details of lithium chemistry can be found in most good chemistry textbooks and in a recently published book dedicated to lithium [9]. In terms of the biochemical behavior of lithium, it is of interest to compare the chemistry of the lithium ion with that of other related cations which are essential for living systems, namely the alkali metal cations, Na^+ and K^+, and the alkaline earth metal cations, Mg^{2+} and Ca^{2+}.

Due to its exceptionally small size, the Li^+ ion is characterized by a charge-radius ratio higher than that of all the other alkali metal cations, and the ion is strongly polarizing, comparable to that of Mg^{2+} and Ca^{2+} (Table 1). Therefore, the chemistry of Li^+ shows many anomalies relative to that of Na^+ and K^+, and often resembles that of Mg^{2+}; the so-called diagonal relationship of Li^+ and Mg^{2+} [10]. The hydrated Li^+ ion has the largest effective diameter, lowest diffusion coefficient, and least lipid solubility of all the alkali metals [11]. Like Na^+ and K^+, Li^+ has a high ionization potential and its salts are generally water soluble. The salts of Li^+ with small anions are exceptionally stable due to the very high lattice energies, whereas those with larger anions are relatively unstable due to poor packing in the crystal. Thus, the solubilities of the Li^+ salts resemble those of the corresponding Mg^{2+} salts: the Li^+ and Mg^{2+} salts of F^-, OH^-,

Table 1. Comparison of Some Chemical and Physical Properties of Lithium with Closely Related Elements

Element	*Atomic Radius (Å)*	*Ionic Radius (Å) (4-coordination)*	*Electro-negativity*	*Polarizing Power (charge/radius²)*
Li	1.52	0.86	0.98	1.88
Na	1.54	1.12	0.93	0.78
K	2.27	1.44	0.82	0.44
Mg	1.60	0.78	1.31	3.97
Ca	1.97	1.06	1.00	1.54

PO_4^-, and CO_3^{2-} are rather insoluble relative to other alkali metal cations, whereas those of ClO_4^-, NO_3^-, and SO_4^{2-} are very soluble.

In biological systems, therefore, the behavior of Li^+ is predicted to be similar to that of Na^+ and K^+ in some cases, and to that of Mg^{2+} and Ca^{2+} in others [12]. Indeed, research has demonstrated numerous systems in which one or more of these cations is normally intrinsically involved, including ion transport pathways and enzyme activities, in which Li^+ has mimicked the actions of these cations, sometimes producing inhibitory or stimulatory effects. For example, Li^+ can replace Na^+ in the ATP-dependent system which controls the transport of Na^+ through the endoplasmic reticulum; Li^+ inhibits the activity of some Mg^{2+}-dependent enzymes *in vitro*, such as pyruvate kinase and inositol monophosphate phosphatase; Li^+ affects the activity of some Ca^{2+}-dependent enzymes—it increases the levels of activated Ca^{2+}–ATPase in human erythrocyte membranes *ex vivo* and inhibits tryptophan hydroxylase.

One model of an "ionic" mechanism of action of Li^+ in affective disorders has been proposed, in which the "receptors" for Li^+ are ion channels and cation coenzyme receptor sites, and in which the presence of intracellular Li^+ in excitable cells results in the displacement of exogenous Na^+ and/or other intracellular cations [13]. It has been suggested that this could lead to a decrease in the release of neurotransmitters; alternatively it may be that this intracellular Li^+ is altering a preexisting, disease-related electrolyte imbalance [14]. A number of observations of such imbalances in affective disorders have been made: depression is associated with elevated levels of intracellular Na^+ [15]; retention of Li^+ is observed in manic-depressive patients prior to an episode of mania [16]; and Na^+/K^+ activity is defective during both mania and depression [17].

2.1 Lithium Complexes

All the alkali metal cations require chelation for significant complexation in aqueous solution. Compared to Li^+, the larger alkali metal cations are more likely to bind neutral organic ligands with carbonyl, ether, or alcohol groups (macrocyclic polyethers, cryptates, and peptides) as these ligands are generally too bulky to fit around the smaller cations; Mg^{2+} and Ca^{2+} bind largely to carboxylate and phosphate anions. The affinity of the cyclic multidentate ligands for a particular ion is strongly dependent upon how well the ion fits into the cavity of that ligand, thus the cationic radius is a critical factor for chelation. A significant amount of research has produced macrocyclic ligands, with polar interiors and hydrophobic exteriors, whose cavity is selective for particular cation sizes, and a substantial number of macrocyclic compounds that are selective for Li^+ can be found in the literature. Li^+ is strongly hydrated in aqueous media and, therefore, these Li^+-selective ligands must be able to remove the shell of water molecules from the ion before it is chelated. These molecules are soluble in organic solvents and can be used to isolate Li^+ from aqueous solutions containing other cations. Two such ligands are the macrocyclic polyether, 12-crown-4 [18], and the macrobicyclic diamine, [2.1.1.] cryptand [19], shown in Figure 1. As cryptands are more structurally rigid than the monocyclic ligands, they tend to show improved selectivity and higher binding affinities. For instance, the [2.1.1] cryptand is perfectly suitable for extracting Li^+ as its cavity diameter is 1.6 Å and it can therefore selectively form a complex with the smaller cation, Li^+, but not with Na^+, which is too large. Ligands which undergo a color change on complexation with ions are also being investigated for use in concentration determinations. The nitrophenyl azo-derivative of 12-crown-4, shown in Figure 1c, is highly Li^+ selective and displays a hypsochromic shift of λ_{max} from 575 to 517 nm on complexation with the cation [20].

Macrocyclic compounds with ion-chelating properties occur naturally and often function as ionophores, translocating ions across biological membranes; many of these compounds are small cyclic polypeptides. Some natural carboxylic polyethers are selective for Li^+ and are, therefore, ionophores for Li^+. Monensin, shown in Figure 1d, is a natural ionophore for Na^+ but it will also complex with Li^+ and it has been shown to mediate the transport of Li^+ across phospholipid bilayers [21]. It has been proposed that synthetic Li^+-specific ionophores have a potential role as adjuvants in lithium therapy, the aim being to reduce the amount of

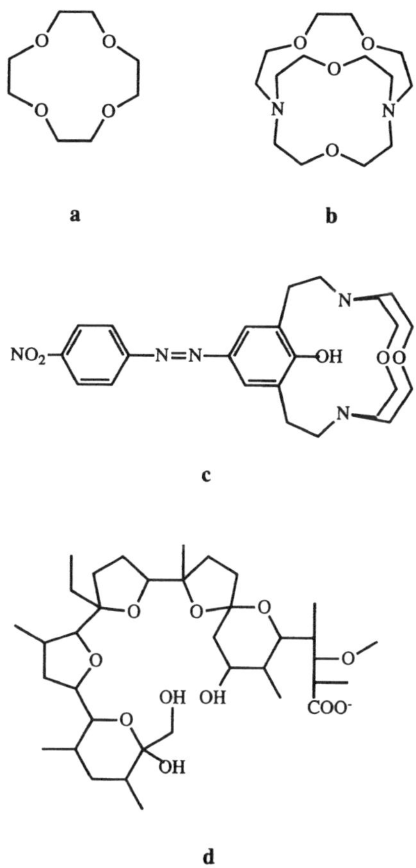

Figure 1. Examples of lithium chelating molecules: (**a**) 12-crown-4 ether, (**b**) [2.1.1.] cryptand, (**c**) nitrophenyl azo-derivative of 12-crown-4, (**d**) monensin.

Li^+ required for efficacy by increasing its bioavailability and possibly to reduce some of the unpleasant side effects sometimes experienced with this drug. The toxicity of such ionophores and kinetics of ion complexation, both association and dissociation, are obviously important factors under consideration in this field of research.

2.2 Analysis in Biological Materials

The analytical techniques employed for the determination of Li^+ in biology have been reviewed in detail [22]. Since Li^+ has no convenient radioisotope, the clinical analysis of Li^+ is primarily achieved using either atomic absorption spectrometry (AAS) or flame emission spectrometry (FES). Many of these analytical methods are destructive to the sample, although a few do have the capability of measuring Li^+ levels *in vivo*, and *ex vivo* nondestructively, such as nuclear magnetic resonance (NMR) spectroscopy and magnetic resonance imaging (MRI). Apart from the obvious advantage of *in vivo* methods, one of the major problems of *ex vivo* analyses arises because the Li^+ ions are very mobile and it is therefore difficult to prepare samples to accurately determine levels in different biological compartments. Many techniques also require very careful calibrations to take into account the significant interference effects from other ions present. Of particular significance is interference by Na^+ which is generally in much higher concentrations than Li^+ in the biological samples. Na^+ is present at approximately 140 mM compared with typically 1 mM Li^+ in blood at therapeutic lithium levels. Some techniques—e.g. NMR spectroscopy, mass spectrometry, and AAS—are also able to discriminate between the two naturally occurring stable isotopes of lithium, ^6Li and ^7Li. The relative mass difference between ^6Li and ^7Li is over 15% and ^6Li has higher charge-to-mass, and charge-to-radius ratios. Therefore, isotopic effects are expected to occur in biology; for instance the electrostatic interactions of the isotopes with water and cell membranes. In erythrocytes, the rate of uptake of ^6Li has been shown to be faster than that of ^7Li when analyzed by either AAS [23] or NMR spectroscopy [24], and a 50% difference in the rate of uptake of the two Li^+ isotopes by rat cerebral cortex has also been observed [25].

As discussed above, Li^+ can be isolated from a solution containing other cations by complexation with specific macrocyclic ligands and subsequent extraction of the Li^+–ligand complex into organic solvents. Complexation of the ligand alters the absorption profile in the UV-vis region and can therefore be followed spectrophotometrically.

Probably the most accessible techniques employed for Li^+ analyses are AAS and FES [26]. Although both of these methods are destructive to the sample and are subject to significant interference effects, the methods have been developed and used successfully for many years. Li^+ levels in solution, in body fluids, and in solubilized tissues have been determined, making a significant contribution to the understanding of Li^+ distribution in the body, and of the membrane transport of Li^+ in various systems.

Regular monitoring of the level of Li^+ in the blood of lithium-treated patients is required as it has to be maintained in the narrow concentration range of approximately 0.3–1.0 mM in order to achieve the acceptable balance between efficacy and toxicity. AAS and FES are the most widely used methods; however Li^+-specific electrodes are playing an increasingly important role in some psychiatric clinics [27]. Since these ion-selective electrodes (ISE) contain a liquid ion-sensitive membrane composed of specific ionophores connected to an electrical circuit, their efficiency obviously depends upon the specificity of the membrane ionophore [28]. In some well-equipped psychiatric clinics, these electrodes are employed to produce rapid determination of the Li^+ level in a patient's blood, allowing any adjustment of the dose to be made immediately and thus reducing the number of patient visits required. Additionally in the laboratory, ISE's with very small, sharp tips can be inserted into living cells for *ex vivo* analyses. For example, Li^+-specific ISE's have been employed to measure the Li^+ content of snail neurones [29].

NMR spectroscopy is a noninvasive, nuclear-specific technique that has been successfully employed to observe and quantify Li^+ *in vivo*, *ex vivo*, and *in vitro*. Li^+ can be observed in different biological compartments simultaneously; for example in intra- and extracellular spaces, by the use of certain chemical reagents which distinguish the responses for Li^+ in each environment. Thus nondestructive investigations of intracellular Li^+ content and membrane transport properties can be achieved. For example, the presence of the paramagnetic shift reagent dysprosium tripolyphosphate, $Dy(P_3O_{10})_2^{7-}$ (Figure 2), in the incubating buffer can result in the separation of the intra-, and the extracellular Li^+ signals in the NMR spectra obtained from human erythrocytes [30], human astrocytoma cells [31], rat hepatocytes [32], and in perfused frog heart [33]. Figure 3 shows typical ^7Li NMR spectra obtained from an experiment to monitor the uptake of Li^+ by human erythrocytes, as evidenced by the increase in the size of the intracellular Li^+ signal as a function of time [34]. Unfortunately, these shift reagents are very toxic and therefore cannot be employed for *in vivo* spectroscopy. Initial attempts to distinguish intra-, and extracellular Li^+ on the basis of other NMR parameters to permit *in vivo* measurements, such as relaxation times or diffusion coefficients, have so far proved unfruitful. Nevertheless, *in vivo* ^7Li NMR spectroscopy has been successfully used to measure both Li^+ levels and uptake of Li^+ in skeletal muscle and in the brain of living animals [35,36] and humans [37–40], and ^7Li MRI has been employed to show the distribution of Li^+ in animal brain [36,41].

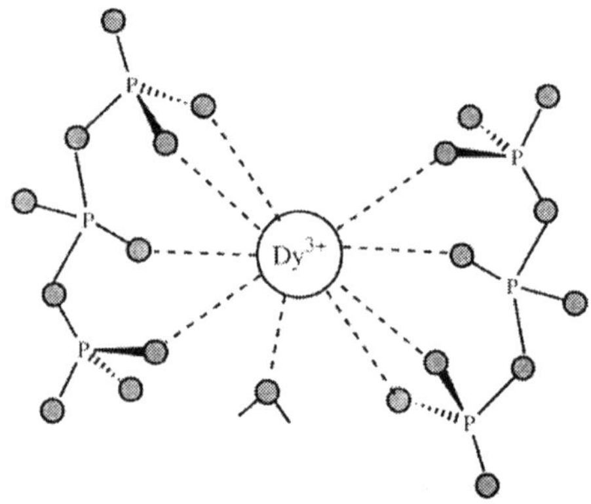

Figure 2. $Dy(P_3O_{10})_2^{7-}$ is a lanthanide shift reagent commonly used in biological ^7Li NMR experiments. The Dy^{3+} ion has a coordination number of nine with two $P_3O_{10}^{5-}$ moieties, acting as tetradentate ligands, and one molecule of H_2O coordinated in the first coordination sphere; up to seven Li^+ ions can bind in the second coordination sphere.

3. BIOLOGICAL DISTRIBUTION

During lithium therapy, the Li^+ cation is widely but unevenly distributed throughout the tissues and fluids, both intra- and extracellular, of the human body. Since Li^+ is very toxic if levels in the body become too high, the concentration of Li^+ in the blood plasma of patients taking the drug has to be carefully monitored and is generally maintained within the range 0.3–1.0 mM by adjustment of the dose. In the brain, the average Li^+ concentration was originally thought to be approximately the same as that in the plasma; however recent studies of psychiatric patients, using *in vivo* magnetic resonance techniques, suggest that Li^+ levels in the brain and muscle are lower than that in the serum [39]. In saliva, the concentration of Li^+ is about twice as high, and in the cerebrospinal fluid it is much lower than in plasma [42]. Studies on animals have shown higher levels of Li^+ in the kidneys, bone [43], and endocrine glands, especially the thyroid [44], and lower levels in the liver [45] and erythrocytes [46] than in serum.

In a study of psychiatric patients after 1 week of lithium treatment, the serum Li^+ level was typically 1 mM, whereas in brain and muscle the levels were 0.4 and 0.5 mM, respectively. Within the brain itself, the distribution of Li^+ appears to be uneven; however no particular region appears to accumulate Li^+ to any significant extent [47]. It has been

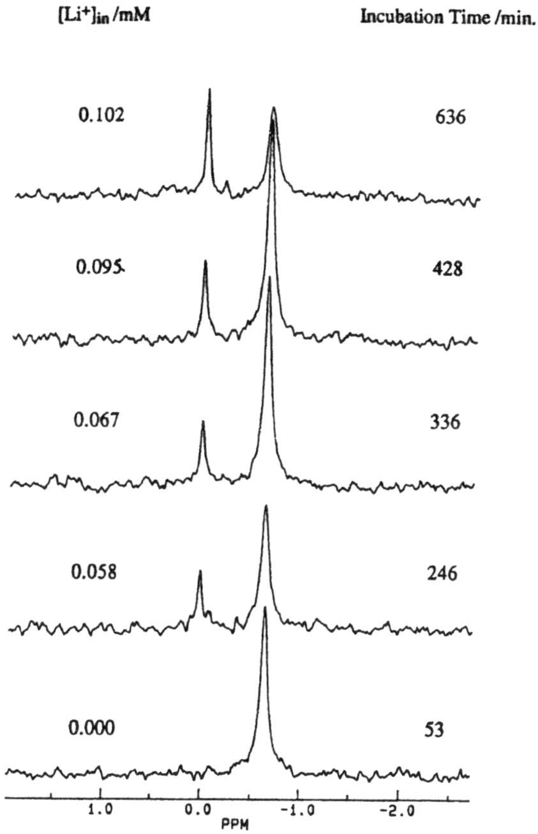

Figure 3. Monitoring the uptake of Li^+ into human erythrocytes after incubation in media containing 2 mM Li^+ using ^7Li NMR spectroscopy. The signals corresponding to the intra-, and extracellular Li^+ are separated by the presence of the paramagnetic shift reagent, $Dy(P_3O_{10})_2^{7-}$, the extracellular medium [34].

reported that the concentration of Li^+ is higher in the pons than in the cerebral white or gray tissue or in cerebellar tissue [48]. The distribution of $^6Li^+$ in a section of mouse brain has recently been imaged by a neutron irradiation technique [49]. This clearly shows a variation in Li^+ accumulation with, for example, higher Li^+ levels in both the hippocampus and hypothalamus, and a lower level in the thalamus. Localized Li NMR and Li-MRI techniques have also been applied *in vivo* to animals and, more recently, to humans. Using these methods it appears that Li^+ is fairly evenly distributed within the brain with no significantly elevated levels being observed in any particular regions [36].

The distribution of Li^+ *in vivo* is primarily due to the relative rates of entry and efflux of the cation in the different tissues. The uptake of Li^+ from the blood is relatively rapid into the kidney and is slower into the liver, bone, and muscle. The movement of Li^+ both into and out of the brain is very slow compared to other organs and this is thought to be due to the low permeability of the blood-brain barrier for this cation [50].

Following oral administration, the intestinal absorption of Li^+ occurs primarily in the small intestine and the subsequent movement of Li^+ into the blood stream is a passive process, via a paracellular route, with very little Li^+ accumulating in the intestinal cells [51,52]. The excretion of Li^+ is almost entirely by the kidneys with only very small amounts (<1%) being excreted in the feces, sputum, sperm, and sweat. The elimination half-life of Li^+ is approximately 20–30 hours.

Most human cells are exposed to less than 2 mM Li^+, and in most tissues the intracellular Li^+ concentration is lower than the extracellular concentration. The level inside cells is generally below that expected for the passive diffusion of the Li^+ ion across the cell membrane, indicating that Li^+ is actively transported out of cells. For instance, the concentration of Li^+ inside the erythrocytes from people taking lithium salts is low with a typical ratio of intra- to extracellular Li^+ of 0.5 [53].

The transport behavior of Li^+ across membranes has been the focus of numerous studies, the bulk of which have concentrated upon the human erythrocyte for which the Li^+ transport pathways have been elucidated and are summarized below. The movement of Li^+ across cell membranes is mediated by transport systems which normally transport other ions, therefore the normal intracellular and subcellular electrolyte balance is likely to be disturbed by this extra cation. Additionally, Li^+ has been shown to increase membrane phospholipid unsaturation in rat brain, leading to enhanced fluidity in the membrane, which could have repercussions for membrane-associated proteins and for membrane transport properties.

Abnormalities in the movement of Li^+ across the erythrocyte membrane have been related to psychiatric disorders and also in response to lithium therapy itself. As yet there is relatively little definitive information about the Li^+ transport mechanisms operating in therapeutically relevant cell types.

3.1 Membrane Transport in Erythrocytes

There are five distinct pathways through which the movement of Li^+ is mediated in human erythrocytes which have been identified and characterized—recently reviewed in Gani et al. [54]. Under artificial conditions of ion concentration, temperature, and pH, Li^+ can be transported into, or out of erythrocytes by any of these mechanisms. However under physiological conditions, the uptake of Li^+ is controlled by just two diffusion processes which are similar to pathways normally transporting Na^+: a concentration-dependent passive transport referred to as a "leak" [55], and a bicarbonate-dependent anion-exchange process which is dependent upon the extracellular concentrations of both Li^+ and bicarbonate ions [56,57]. The molecular carrier for the latter system is a component of the band 3 protein in the erythrocyte membrane. The low intracellular concentration of Li^+ is achieved by the extrusion of Li^+ mediated by an ATP-independent, Na^+–Li^+ countertransport process in which one Li^+ ion is transported out of the cell as one Na^+ ion enters [58, 59]. This transport mechanism is driven by the transmembrane Na^+ concentration gradient and it is believed to be the same as the Na^+–Na^+ exchange system which is found in erythrocyte membranes. This process observes Michaelis–Menten kinetics with a K_m for Li^+ of approximately 0.5 mM at the internal side of the membrane [60]. It has a higher affinity for Li^+ than Na^+ and has a greater affinity for both cations at the internal side of the erythrocyte membrane than the external. Under physiological conditions of relatively high extracellular and low intracellular concentrations of Na^+, the Li^+ competes effectively with the intracellular Na^+ and is transported out of the cell. Physiological concentrations of intracellular Na^+ and extracellular K^+ inhibit the transport of Li^+ by either of the two remaining pathways which are capable of transporting Li^+: the Na^+–K^+ pump and the Na^+–K^+ cotransport system [56,61].

3.2 Variability in Transport in Erythrocytes

The ratio, $[Li^+_{in}]/[Li^+_{out}]$, in human erythrocytes ranges from 0.2–0.9, illustrating a large inter-individual variability; however this ratio is relatively constant over long periods of time for individuals and there is evidence that this phenomena is under genetic control [62]. Higher ratios have been reported in patients with affective illnesses compared to healthy subjects [63] and in females compared to males [64]. There have been attempts to correlate the higher Li^+ ratio to a better clinical response of patients to Li^+ therapy [65].

It has been shown that these variations in the steady-state Li^+ ratios appear to be due primarily to differences in the Na^+–Li^+ countertransport mechanism; a higher Li^+ ratio arising almost exclusively from a lower activity of this mechanism [56]. Significant individual variation in the maximal transport activity, V_{max}, has been reported by a number of groups; Sarkadi reported a range of 0.1–0.37 mmol Li^+(l cells)$^{-1}$ h^{-1} [60]. Interestingly, this transport process appears to show little variability over time within an individual, indicating that the magnitude of this transport process is a characteristic property of the erythrocyte membranes of an individual and thus suggesting the involvement of a genetic component [66].

Differences in the Na^+–Li^+ countertransport process have also been observed in the erythrocytes of psychiatric patients [67]. Pandey reported that this process was generally slower in bipolar patients than in normal controls, leading to the observed higher Li^+ ratios, although many patients still had Li^+ ratios in the "normal" range [68].

A decrease in the efficiency of the Na^+–Li^+ countertransport is also observed as a direct result of Li^+ administration, with a 50% inhibition in the efflux of Li^+ from the erythrocytes of people on Li^+ therapy [63]. This decrease in activity occurs 2–4 days after commencing therapy and maximum reduction appears within 7 days; the rate of transport returns to normal soon after the Li^+ administration is ceased. This Li^+-induced change has been attributed to a decrease in the affinity of the transporter for Li^+ as the K_m for the process increases threefold, whereas V_{max} remains constant, in contrast to the interindividual variability [69].

Variations in the Na^+–Li^+ countertransport transport mechanism in other mammalian species, namely sheep, beef, horses, and rabbits, have also been observed [70]. Here, as in the cases of the interindividual differences, the maximal transport activity, V_{max}, is altered, being lowest in humans and significantly higher in the others. Interestingly, there is no corresponding Na^+–Li^+ countertransport process in rat erythrocytes, and Li^+ transport via the "leak" route is approximately threefold greater in rat erythrocytes than in humans [70]. This results in higher levels of Li^+ being attained in rat erythrocytes compared to other mammals as demonstrated in a recent ^{7}Li NMR study (Figure 4) [34].

3.3 Membrane Transport in Other Cells

Erythrocytes share some transport mechanisms with many cell types although, surprisingly, the Li^+ transport pathways in other cells have not

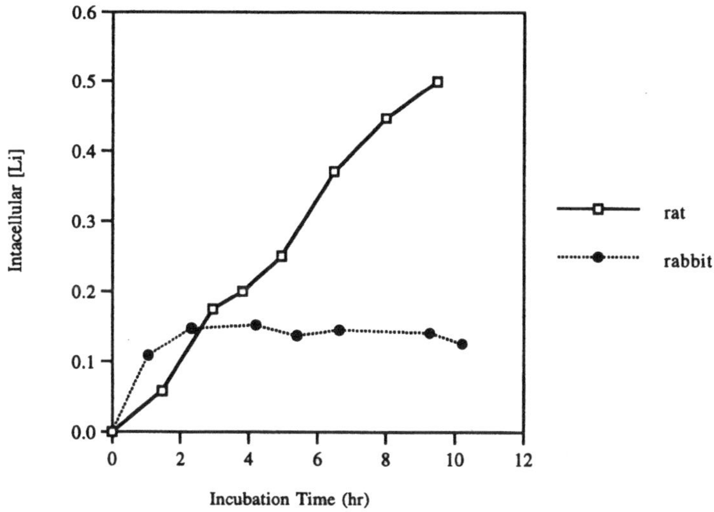

Figure 4. Comparison of the uptake of Li^+ into erythrocytes from rat and rabbit, determined by ^7Li NMR spectroscopy [34].

been thoroughly investigated. Li^+ does not appear to be taken up into cells to any great extent as in erythrocytes; however there are some exceptions which may prove to be therapeutically significant.

Many cells, including those in the brain and in muscle, exhibit a Na^+–Na^+ exchange system similar to that observed in human erythrocytes [55]; however the majority of cells lack the anion transporter present in erythrocytes. Neurones also possess a voltage-dependent Na^+ channel which is not found in nonexcitable cells and the permeability of this transporter to Li^+ is almost the same as that to Na^+. As with Na^+, the cell membrane becomes dramatically more permeable to Li^+ in the excited state with, for instance, veratridine which is an alkaloid which activates these Na^+ channels, and this significantly increases the initial rate and the extent of Li^+ uptake in mouse neuroblastoma cells [71] and in neuroblastoma x glioma hybrid cells [72]. Li^+ ratios of 5.6, 1.5, 0.3, and 0.7 have been observed in rat glioma cells [73], mouse neuroblastoma cells [71], mouse neuroblastoma/glioma hybrid cells [72,74], and primary chick embryo brain containing mixed populations of neuronal and glial cells [75,76], respectively. In neuroblastoma/glioma cells the movement of Li^+ is mediated in part by the voltage-dependent Na^+ channel and in part by a Li^+–Na^+ countertransport mechanism similar to that operating in erythrocytes [74]; a further pathway transporting Li^+ out of the cell against its concentration gradient has also been identified but not characterized. A study on human neuroblastoma and glioma cell lines shows that Li^+ movement appears to be mediated by the Na^+–K^+ pump in the neuroblastoma but not in the glioma cells, and there is no evidence for a Li^+–Na^+ countertransport pathway; the glioma cells attained a significantly higher level of Li^+ [77]. In general, these cells have faster transport rates than those observed for erythrocytes since the intracellular Li^+ achieves steady-state within 20–60 minutes compared to several hours for erythrocytes. Variability in intracellular Li^+ concentrations across cell types is the expected consequence of several transport pathways operating at different levels, and probably accounts for the observed variability in the degrees of the Li^+-induced responses among the different cell types and tissues.

In the proposed ionic mechanism of action of Li^+ [13], it is assumed that Li^+ entry via the voltage-dependent Na^+ channel displaces Na^+ from excitable cells, thus decreasing the Na^+–Ca^{2+} countertransport at the synapse leading to a decrease in the Ca^{2+} transient and the resulting release of neurotransmitters. Since Li^+ entry is proposed to be primarily through this voltage-dependent pathway, this response will be significant in active neurons which are likely to be stimulated during the extreme mood swings of manic depression, whereas during normal mood states these neurons will be less active and the entry of Li^+ will be much reduced. Thus, the therapeutic effect of Li^+ in the proposed model arises from its ability to act on abnormally stimulated rather than on resting neurones.

4. PHOSPHOINOSITIDE METABOLISM

The most popular hypothesis focused upon in recent years for the therapeutic effect of Li^+ is the significant interference of this ion with the inositol phospholipid-dependent intracellular signaling system [78]. Many hormones, neurotransmitters, and growth factors activate the enzyme phospholipase C (PLC), primarily resulting in the activation of two second messengers: inositol 1,4,5-trisphosphate (Ins(1,4,5)P_3) and *sn*-1,2-diacylglycerol (DAG) [79]. Li^+ interferes with inositol lipid metabolism by inhibiting the intracellular enzyme, inositol monophosphate phosphatase (IMPase), which is involved in the crucial recycling of the inositol phosphates.

Since the effects of Li^+ on this system are primarily observed in cells that are actively stimulated, the hypothesis is that Li^+ alleviates manic-depression by interfering with the transduction of aberrant inositol phospholipid-dependent signals in abnormally stimulated neurones, such as would arise in the manic or depressive phases of the illness [80]. During episodes of "normality," when the neurones are experiencing less intense stimulation, the transduction of these signals occurs to a lesser extent and Li^+ has limited influence on the phosphoinositide metabolism. This correlates with the fact that Li^+ has significant tempering effects on both the depressive and the manic phases of manic-depression while having little effect on the normal mental functions of patients.

4.1 Li^+ Effects upon the Phosphoinositide Cycle

A simplified scheme of the phosphoinositide cycle, as it is understood today, is illustrated in Figure 5 and is explained in detail elsewhere [81]. The activation of specific cell-surface neurotransmitter receptors is coupled to the stimulation of PLC via a guanine nucleotide-binding protein (G protein). This enzyme then catalyzes the hydrolysis of the membrane phospholipid, phosphatidylinositol 4,5-bisphosphate (PtdIns(4,5)P_2), producing the two intracellular second messengers, Ins(1,4,5)P_3, which facilitates the release of intracellular Ca^{2+} into the cytosol from intracellular stores, and DAG, which activates and facilitates the translocation of protein kinase C (PKC) from the cytosol to the membrane [79]. As shown in Figure 5, Ins(1,4,5)P_3 is metabolized either by sequential dephosphorylation to inositol, via Ins(1,4)P_2 and Ins(4)P, or phosphorylated to Ins(1,3,4,5)P_4 followed by sequential dephosphorylation, via either Ins(1)P or Ins(3)P. The relative rates of the different routes for the degradation of Ins(1,4,5)P_3 are tissue-dependent; for instance in cerebral cortical slices, Ins(1,4,5)P_3 is metabolized predominantly via Ins(1,3,4,5)P_4 resulting in accumulation of Ins(1)P rather than Ins(4)P [82]. Inositol is recycled as shown and it is believed that several other inositol polyphosphates in this cycle have important, possibly cell-signaling, functions [83]. PtdIns(4,5)P_2 is also a substrate for another receptor-coupled enzyme, phosphoinositide 3-kinase (PI 3-kinase) which results in the formation of another potential second messenger, phosphatidylinositol 3,4,5-trisphosphate (PtdIns(3,4,5)P_3) [84].

The relatively small pool of the precursor for Ins(1,4,5)P_3, the phospholipid PtdIns(4,5)P_2, is synthesized from the larger reservoir of phosphatidylinositol (PtdIns) and, therefore, the supply of PtdIns(4,5)P_2

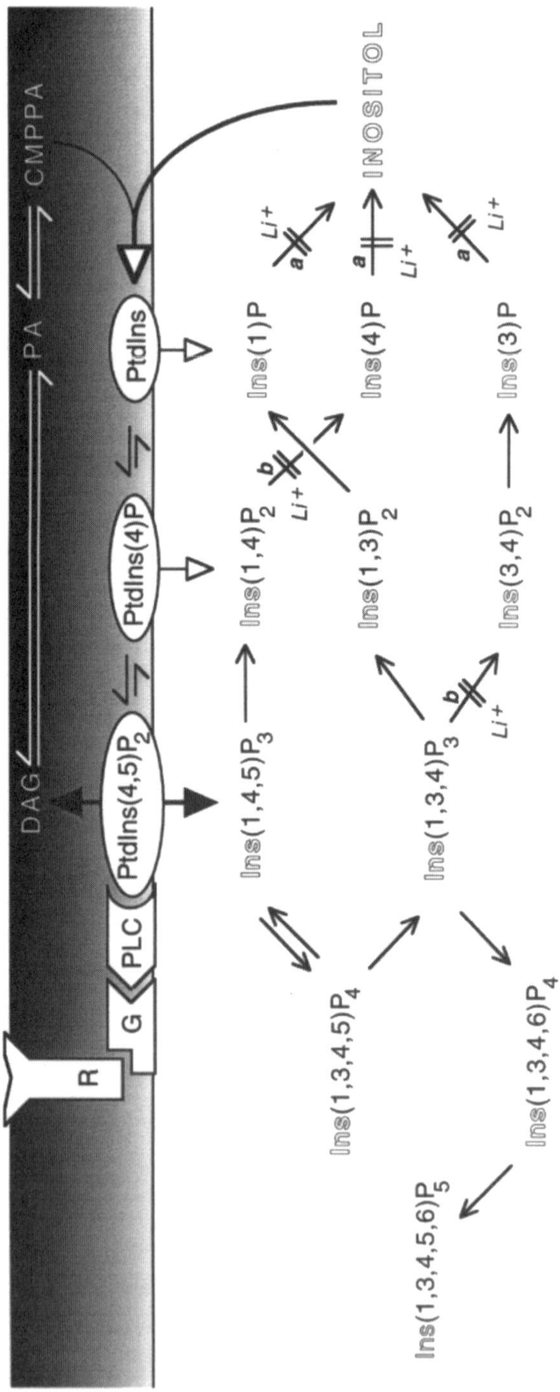

INOSITOL
CMPPA
PA
DAG
PtdIns
PtdIns(4)P
PtdIns(4,5)P2
PLC
G
R
Ins(1)P
Ins(4)P
Ins(3)P
Ins(1,4)P2
Ins(1,3)P2
Ins(3,4)P2
Ins(1,4,5)P3
Ins(1,3,4)P3
Ins(1,3,4,5)P4
Ins(1,3,4,6)P4
Ins(1,3,4,5,6)P5
a
b
Li+

available to the receptor depends upon maintenance of the levels of PtdIns and hence the levels of its precursor, cellular inositol. The three sources which contribute to the pool of free inositol in cells are: the phosphoinositide cycle, whereby the final step is dephosphorylation of the inositol monophosphates; *de novo* synthesis from glucose, in which the penultimate step is the cyclization of glucose-6-phosphate to Ins(3)P, which is then dephosphorylated; and uptake from the extracellular environment. Li^+ inhibits both the first two of these pathways and, thus, it has been postulated that the recycling of PtdIns is likely to be significantly affected by this cation, particularly in the brain which is thought to have a limited access to peripheral supplies of extracellular inositol [78,80].

The profound effects of Li^+ upon phosphoinositide metabolism and cell signaling have been the subject of several recent reviews [54,81,85,86]. These effects are dependent upon receptor stimulation of the phosphoinositide cycle by a range of stimuli, including norepinephrine, serotonin, and carbachol; the basal turnover of this cycle is largely unaffected by Li^+ [82,87,88].

Li^+ was first found to interfere with inositol lipid metabolism when significantly decreased levels of *myo*-inositol were observed in the cerebral cortex of Li^+-treated rats [89]. Subsequent work revealed a corresponding increase in the levels of Ins(1)P [90] and this behavior was shown to be the result of a Li^+-induced inhibition of IMPase, the enzyme which dephosphorylates the monophosphates Ins(1)P, Ins(3)P, and Ins(4)P to produce free inositol [91]. These results stimulated much research in this field involving a wide variety of cell types, tissues, and animals where the Li^+ inhibition of IMPase was found to be ubiquitous. However, it was found that, *in vivo*, this Li^+-induced effect is predominantly limited to the brain, being observed in different regions of the brain to different extents, with similar results for both acute and chronic treatment with Li^+. It is probable that those cells that are able to accumulate inositol, or which are exposed to and can rapidly import an extracellular supply of inositol, may be relatively insensitive to the effects of Li^+.

The Li^+-induced inhibition of IMPase also results in increases in the levels of the other monophosphates, Ins(3)P, and Ins(4)P, although the increase is less than for Ins(1)P in the brain. Increases in the levels of the bisphosphates are also observed and these increases are believed to be related to the Li^+-induced inhibition of another enzyme in this cycle, inositol polyphosphate 1-phosphatase [92], which dephosphorylates both Ins(1,3,4)P_3 and Ins(1,4)P_2, producing Ins(3,4)P_2 and Ins(4)P, respectively. Under different conditions of extracellular inositol levels, both increases and decreases in the levels of the trisphosphates, Ins(1,3,4)P_3 and Ins(1,4,5)P_3 [93,94], have been observed. Li^+ also causes a delayed reduction in the carbachol-induced increase in Ins(1,3,4,5)P_4 although the mechanism, or consequence, of this latter effect is, as yet, unknown [88].

The decreased level of free inositol leads to a reduction in the rate of PtdIns resynthesis and to the subsequent accumulation of cytidine monophosphorylphosphatidate (CMP-PA), the cosubstrate for the resynthesis, and of the other lipid metabolites, phosphatidic acid (PA) and DAG [95]. The short-term increase in DAG levels following acute Li^+ administration is proposed to account for the subsequent enhanced activation of PKC, facilitating the release of certain neurotransmitters. PKC undergoes rapid proteolysis after its activation and it is proposed that chronic Li^+ treatment, producing prolonged increases in DAG levels, actually downregulates PKC activity and attenuates neurotransmitter release [96]. This is reflected in the observed increase in the release of serotonin in rat hippocampus with acute Li^+ administration, followed by decreased release of serotonin after chronic exposure to Li^+ [97,98].

After prolonged stimulation, the pools of inositol phospholipids eventually become depleted to the extent that the rate of synthesis of Ins(1,4,5)P_3 is reduced [93,99], although whether this is sufficient to affect the PI 3-kinase pathway is not yet known. Not all the observed Li^+-induced effects on the phosphoinositide system are in accord with the above hypothesis. For instance, the levels of the inositol phospholipids, PtdIns, PtdIns(4)P, and PtdIns(4,5)P_2 in samples of whole brain are surprisingly not significantly affected by the depletion of inositol [100]. It has been suggested that the lack of significant reduction in the levels of the inositol phospholipids in whole brain infers that the influence of Li^+ on this second-messenger system is less significant than originally hypothesized. However, it may be that the Li^+-induced effects are localized to a small population of hyperstimulated neurones and that these might not be reflected in the analysis of the whole brain [101]. It has also been shown that resynthesis of the inositol phospholipids may occur by base-exchange reactions of other phospholipids, for instance, phosphatidylcholine which is present at much higher levels than PtdIns; however whether this is quantitatively significant is not yet known [96,102].

It was originally thought that the blood–brain barrier is relatively impermeable to inositol and, in consequence, neurones are primarily dependent upon the Li^+-sensitive recycling pathways for supplies of free inositol; however this is now questionable [103]. As a result of the implied Li^+-induced reduction of inositol levels, it was originally proposed that neurones may be specifically vulnerable to the inhibitory action of Li^+, particularly during periods of intense stimulation, in comparison to peripheral cells which are exposed to supplies of inositol in plasma [78,104]. This behavior has two interesting aspects in relation to the therapeutic action of Li^+. First, the time delay before a response to Li^+ treatment is observed is accounted for by the reservoirs of PtdIns and inositol available in cells, which have to be depleted before the inhibitory effect of Li^+ becomes significant. Second, the inhibitory effect of Li^+ will be significant only in cells when the receptors are being hyperstimulated, as might be expected in neurones during those periods of severe mania or depression. In this respect, the uncompetitive mode of inhibition of IMPase by Li^+ (see below) means that the Li^+ effect is considerably more significant when the enzyme is stimulated compared to basal turnover. Additionally, differential inositol transport behavior between cell types has been observed: some cell types can transport inositol rapidly while others lack active inositol transport pathways, and some cells have the ability to accumulate large reservoirs of inositol. Therefore, while some cells will be relatively protected from, others will be selectively vulnerable to the effects of Li^+ upon phospholipid metabolism. This has significant bearing on the specificity of Li^+ in its therapeutic actions in that, although the Li^+-induced inhibition of inositol monophosphatase has been observed in all tissues in which it has been studied, the consequential effect upon the cell signaling mechanisms may be limited to those cells which rely upon Li^+-sensitive pathways for supplies of inositol.

Whatever the therapeutic mechanism of action of Li^+ is, the facts are that Li^+ does indeed inhibit the turnover of the phosphoinositide cycle and does alter the levels of both second messengers, Ins(1,4,5)P_3 and DAG, either or both of which may have wide-reaching secondary effects.

4.2 Inositol Monophosphate Phosphatase and Inositol Polyphosphate 1-Phosphatase

The mechanism of the action of IMPase and its inhibition by Li^+ has been studied in depth, and is detailed by Gani et al. [54]. IMPase catalyzes the hydrolysis of all the *myo*-inositol monophosphates except Ins(2)P. It is completely dependent upon Mg^{2+} for activity; Ca^{2+} and Mn^{2+}

are competitive with Mg^{2+} [91], and inositol 2-phosphate (Ins(2)P) and inorganic phosphate are competitive with Ins(1)P [105]. At therapeutically relevant concentrations of Li^+ (0.5–1.2 mM), this cation acts as an uncompetitive inhibitor with respect to the substrate, Ins(1)P, and is noncompetitive with respect to Mg^{2+}, with reported K_i's of approximately 0.8 mM for the hydrolysis of Ins(1)P and 0.26 mM for that of Ins(4)P [91,106–108]. Also, it has been shown that high concentrations of Mg^{2+} also inhibit this enzyme uncompetitively with respect to Ins(1)P [105] and that inhibition by fluoride has a similar profile to that by Li^+ [109]. At higher concentrations of Li^+, the inhibition of this enzyme is noncompetitive with respect to Mg^{2+} [105]. Interestingly, the inhibitory effect of chronic Li^+ treatment does not appear to be compensated for by an increase in the activity of IMPase [110].

Uncompetitive inhibition is extremely rare in nature, and can arise when the inhibitor binds to the enzyme–substrate complex, rather than to the free enzyme, as in competitive inhibition [111]. In uncompetitive inhibition, an increase in the concentration of the inhibitor requires a disproportionately large increase in the concentration of the substrate to maintain the same metabolic turnover.

IMPases have been purified from bovine brain [91,107] and rat brain [106]. The enzymes from human brain [112] and bovine brain [113] have also been cloned and expressed in and purified from *E. coli*. The enzyme is functionally active as a homodimer comprising two 30 kDa subunits. The three-dimensional structure of the human IMPase complexed with Gd^{3+} and sulfate has been determined by X-ray crystallography [114] and was initially analyzed to identify the potential active site.

Subsequent investigations of the mechanism of the hydrolysis of Ins(1)P have involved detailed chemical kinetics [105,115,116], site-directed mutagenesis [117], fluorescence spectroscopy [118], and analysis of the more recently determined crystallographic structures of the enzyme complexed with Gd^{3+} and Ins(1)P, with Gd^{3+} and Ins(3)P, and with Mn^{2+} [119,120]. These studies show that the hydrolysis of Ins(1)P, catalyzed by IMPase, is complex and appears to involve an ordered ternary mechanism. It is apparent that two separate binding sites for Mg^{2+} are involved in the catalysis and that inhibition of the enzyme by Li^+ is caused by this cation binding at one of these Mg^{2+} binding sites. In the currently accepted mechanism for the hydrolysis, the first Mg^{2+} ion is complexed to the enzyme at a site deep within the substrate binding cleft. This site probably corresponds to that occupied by Gd^{3+} in the crystal structure [119]. Ins(1)P then binds in the substrate binding site, followed by the second Mg^{2+} ion. The second substrate, H_2O, acts as the nucleophile, attacking the ester and hydrolyzing it to produce inositol and inorganic phosphate, P_i. Inositol is then released from the complex first, followed by the second Mg^{2+} ion, and finally P_i. Since the second Mg^{2+} ion binds to the enzyme after Ins(1)P does, the implication is that the substrate either modifies or creates the binding site to accommodate the second Mg^{2+}. Interference of this pathway by Li^+ occurs after the second Mg^{2+} ion is released when a Li^+ ion then binds to the active site which is vacated by this Mg^{2+} ion, before P_i is released, resulting in an enzyme–Mg^{2+}–P_i–Li^+ product complex [115]. In complexation with Li^+ in this manner, IMPase is rendered completely inactive.

At higher concentrations of Li^+, it is proposed that the observed noncompetitive inhibition of IMPase [105] is due to Li^+ binding to the first Mg^{2+} site on the free enzyme [115]. As predicted from the crystal structure, this site is located deep within the active cleft of the enzyme and is therefore relatively inaccessible to the Li^+.

Inositol polyphosphate 1-phosphatase shares several biochemical properties with IMPase: it is also dependent upon Mg^{2+}-for catalytic activity, it is uncompetitively inhibited by Li^+, and it is inhibited by Ca^{2+}. This similarity of behavior suggests that the two enzymes are related in some way. Bovine inositol polyphosphate 1-phosphatase has been cloned [121] and the recombinant enzyme has been expressed in, and purified from baculovirus-infected insect cells [122]. The three-dimensional structures of this enzyme complexed with Mg^{2+}, and with Gd^{3+} and sulfate, have been determined by X-ray crystallography [123]. The enzyme exists as a 44 kDa monomer and, although it is considerably larger than a subunit of IMPase and there is no significant overall primary sequence homology, there are motifs in the sequences which are common to the two enzymes and which appear to involve metal binding and substrate catalysis. The tertiary structures of the two enzymes share a significant degree of similarity (Figure 6) with the proposed active site in similar locations. As with IMPase, there are apparently two Mg^{2+} binding sites, strongly suggesting a similar mechanism for the hydrolysis reaction [123].

5. ADENYLATE CYCLASE-DEPENDENT SIGNALING

The second popular hypothesis for the therapeutic mechanism of action of Li^+ is its interference with another receptor-coupled, second messenger system, that of the plasma-membrane bound enzyme, adenylate

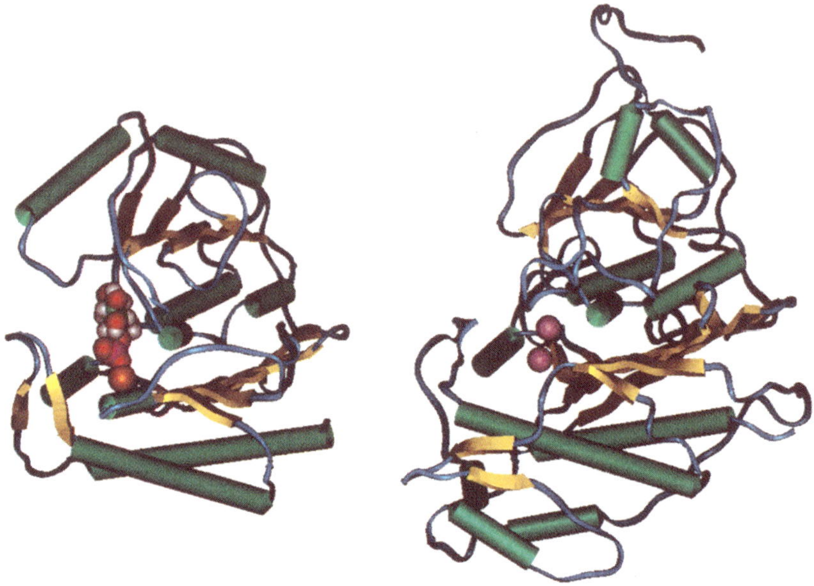

Figure 6. Schematic representation of inositol monophosphate phosphatase (left) and inositol polyphosphate 1-phosphatase (right), showing the helical (green cylinders) and β-sheet (yellow arrows) regions. The monophosphatase is complexed with Ins(1)P (solid spheres) and Gd^{3+} (orange sphere) in the binding cleft and the polyphosphatase has two Mg^{2+} (lilac spheres) ions. (The coordinates were obtained from the Brookhaven Protein Data Bank).

cyclase. This system, which can be either stimulated or inhibited by a number of hormones and neurotransmitters, is analogous to that of the phosphoinositides in that the signal from the agonist-stimulated receptor is coupled to the second messenger-generating enzyme, in this case adenylate cyclase, via a G-protein.

This transmembrane signaling system involves a complex consisting of several functional proteins (Figure 7): stimulatory (e.g. β-adrenergic, dopamine D_1, serotonin, vasopressin) [124] and inhibitory (e.g. α_2-adrenergic, dopamine D_2, opiod, and muscarinic) [125] receptors, stimulatory (G_s) and inhibitory (G_i) G-proteins, and the catalytic protein, adenylate cyclase. On stimulation of a receptor, an associated G-protein binds GTP and the resulting receptor/G-protein/GTP complex then activates, or inhibits, adenylate cyclase in the catalysis of the synthesis

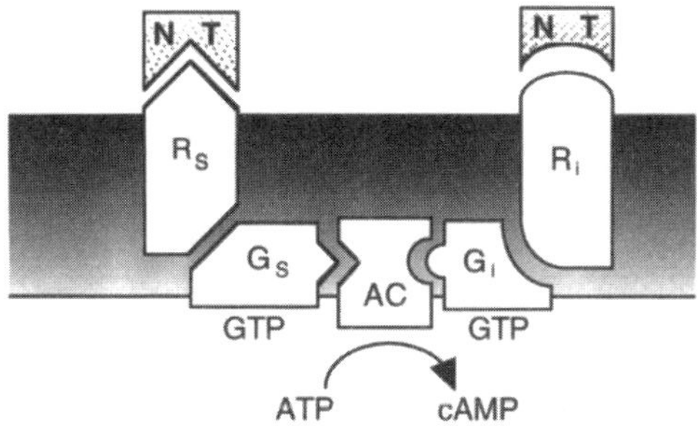

Figure 7. A simplified representation of the receptor adenylate cyclase complex.

of adenosine 3′,5′-phosphate (cAMP) from ATP. cAMP is the intracellular second messenger which goes on to activate a number of cAMP-dependent protein kinases [126].

Adenylate cyclase is dependent upon Mg^{2+}, which complexes with ATP to form the substrate, $MgATP^{2+}$. Mg^{2+} is also essential for the regulation of enzymatic activity, having one binding site on adenylate cyclase and another on G_s, modulating the receptor–agonist affinity and the activity of this G-protein [127]. Protein kinase C (PKC) also modulates the activity of adenylate cyclase although the mechanisms for this are not fully understood. PKC is reported to phosphorylate the enzyme directly [128], enhancing its activity, and to phosphorylate G_i, thereby reducing its activity [129]. Moreover, activated adenylate cyclase is reported to increase the activity of PKC. Adenylate cyclase can be stimulated more directly by activating the G-protein, thus bypassing the receptor, or by direct activation of the enzyme itself, for instance by forskolin [130].

The effects of Li^+ upon this system have been reviewed in depth by Mork [131]. Animal studies originally demonstrated that Li^+ inhibits cAMP formation catalyzed by adenylate cyclase in a dose-dependent manner [132]. The level of cAMP in the urine of manic-depressive patients changes with mental state, being abnormally elevated during the "switch" period between depression and mania; it is proposed that Li^+'s inhibitory effect upon adenylate cyclase activity may correct this abnormality. Subsequent research, in accord with the initial experiments, have shown that Li^+'s interference with this second messenger system involves more than one inhibitory action. At therapeutic levels, Li^+ inhibits cAMP accumulation induced by many neurotransmitters and hormones, both *in vivo* and *in vitro* [133]. As in the phosphoinositide cycle, the basal activity of the adenylate cyclase-dependent second-messenger system is unaffected by Li^+. Thus the hypothesis argues that Li^+ exerts its effect primarily when this system is abnormally stimulated, during the extreme mood swings of manic-depression. The multifactorial influence of Li^+ on this system may also be significant in its therapeutic mechanism of action. The current evidence for the target for Li^+'s action, with respect to the signal transduction pathways of adenylate cyclase and phosphoinositide metabolism, has been reviewed recently [86].

Li^+, at therapeutically relevant concentrations, is a potent inhibitor of norepinephrine-stimulated adenylate cyclase activity *ex vivo* in both rat [133] and human brain [134], and it inhibits norepinephrine-stimulated cAMP accumulation in Li^+-treated patients. Li^+ also inhibits dopamine-stimulated cAMP accumulation in rat brain [135]. These inhibitory effects of Li^+ have been shown to be region specific within rat brain, a fact that has obvious significance for a therapeutic mechanism of action. It is interesting that other antimanic drugs may also have dampening effects on dopaminergic neurotransmission.

Li^+ also inhibits several hormone-stimulated adenylate cyclases which, in some cases, appear to be related to side effects of Li^+ therapy. For instance, Li^+ inhibits the hydro-osmotic action of vasopressin, the antidiuretic hormone which increases water resorption in the kidney [136]. This effect is associated with polyuria, a relatively harmless side effect sometimes experienced with Li^+ treatment, which arises from the inability of the kidney to concentrate urine. Li^+ has been shown to inhibit vasopressin-stimulated adenylate cyclase activity in renal epithelial cells. Additionally, Li^+ is reported to enhance the vasopressin-induced synthesis of prostaglandin E2 (PGE_2) *in vitro* in kidney. PGE_2 inhibits adenylate cyclase activity by stimulation of G_i, and, therefore, this effect may contribute to the Li^+-induced polyuria.

In some cases, the influence of Li^+ on the adenylate cyclase system appears to be localized at the G-protein. For instance, *ex vivo* Li^+ inhibits agonist-induced increases in the binding of GTP to the G-protein, possibly by competing with Mg^{2+} for a binding site on the G-protein [137]. However, the Li^+-induced inhibition of adenylate cyclase by a nonhydrolyzable analogue of GTP, guanyl-5′-imidodiphosphate (Gpp(NH)p)-, and isoprenaline-stimulated adenylate cyclase activity is not overcome by Mg^{2+}. In manic-depressive patients, the affinity for GTP binding, as evidenced by binding of Gpp(NH)p, is increased relative to controls and is consistent with an increased agonist-induced cAMP accumulation in manic-depressives. It has been suggested that this infers an abnormality in the G-protein and, therefore, Li^+'s inhibitory effect upon GTP binding may be involved in its therapeutic effect. Li^+ also inhibits postreceptor adenylate cyclase activity in specific cases [138]: it inhibits GTP-, Gpp(NH)p-, and NaF-stimulated adenylate cyclase activity, all of which act through G_s. There is a lag period in the response of adenylate cyclase to Gpp(NH)p which may reflect the time for G_s to be converted from an inactive to an active form; Li^+ does not affect this lag period *ex vivo* and therefore cannot be interfering with this activation of G_s [139]. Finally, Li^+ inhibits the direct activation of adenylate cyclase, independent of effects at the receptor and the G-proteins: it inhibits both forskolin stimulation, and Ca^{2+}-dependent, calmodulin stimulation of adenylate cyclase activity *ex vivo* in rat brain [130].

Mg^{2+} is competitive with the Li^+ inhibition of both postreceptor G-protein stimulation [140], and direct stimulation of adenylate cyclase [141]. Li^+ inhibits Mn^{2+}-stimulated adenylate cyclase activity in membranes in the presence, but not in the absence, of calmodulin. Since, Mn^{2+} can replace Ca^{2+} in activating calmodulin, it is likely that the observed inhibition is that of the Mn^{2+}-dependent calmodulin stimulation of the enzyme. In the absence of calmodulin, stimulation of adenylate cyclase is probably due to substitution of Mn^{2+} for Mg^{2+} in the substrate, $MnATP^{2+}$ and this is unaffected by Li^+.

It is obvious that Li^+ has several inhibitory effects on the adenylate cyclase-dependent second-messenger system. Many of these influences can be counteracted by the addition of Mg^{2+}, therefore it is probable that Li^+ is competing for Mg^{2+} binding sites on, and thereby interfering with the activity of, the catalytic enzyme itself and/or of the associated G proteins.

6. NEUROTRANSMITTERS AND HORMONES

Numerous reports of altered neurotransmitter and hormone functions which have been associated with the affective disorders are reviewed by Levell [142]. It was originally proposed that one or more of the neurotransmitter amines in the brain (norepinephrine, dopamine, serotonin) may be functionally elevated in manic patients and reduced in depressed patients [143]. For instance, an increase in the production of dopamine, observed in a number of case reports, is thought to be the cause of the switch into the manic phase in bipolar patients. For example, Bunney et al. reported an increase in the level of homovanillic acid (HVA), a metabolite of dopamine, in the CSF just before the switch into mania [144]. Also the pineal gland and its hormone, melatonin, are sometimes implicated with the affective disorders, including the use of bright light in the treatment of seasonal affective disorders. Manic-depressives are also supersensitive to light and there is a case report of increased levels of melatonin during mania and of cortisol during depression [145].

Whatever the mechanism of action of Li^+ in the treatment of bipolar affective disorder turns out to be, there is no doubt that the functions of one or more of the neurotransmitters and hormones are involved to some extent. Much of the published data on the effects of Li^+ on these systems is equivocal or even contradictory, in many cases reflecting differences in the experimental procedures, in particular the levels of Li^+ employed. Often, where it has been looked for, there are differences observed between the acute and chronic effects of Li^+. Therefore, the therapeutic relevance of many of these Li^+-induced effects is difficult to assess.

6.1 Neurotransmitters

The numerous effects of Li^+ upon the neurotransmitters [146,147] and their membrane receptors [148] have been reviewed recently. Li^+ affects processes involved with the synthesis of, release of, reuptake of, and receptor activation by neurotransmitters in both animals and humans. In terms of the neurotransmitter amines, serotonin appears to be most affected by Li^+, whereas the effects upon dopamine and norepinephrine are small.

Much of the data on the effects of Li^+ upon dopamine and its metabolites is contradictory. The turnover and concentrations of dopamine do not appear to be affected by Li^+ in rats, although a Li^+-induced decrease in the number of dopamine receptors in rat striatum has been reported [149]. Increased levels of HVA, a metabolite of dopamine, have been observed in the CSF of patients on Li^+ [150].

The metabolism of norepinephrine is reported to be altered by other drugs used in the treatment of the affective disorders and a number of studies have shown a change in the metabolism of norepinephrine as a result of Li^+ treatment. In rat brain, acute Li^+ treatment enhances the uptake of norepinephrine in synaptosomes [151] and the enhanced turnover of this neurotransmitter may be due to an increase in its deamination in the brain, although Li^+ also causes a slight increase in the levels of the amino acid precursor, tyrosine, in the brain and plasma of rats [152]. Also, acute Li^+ treatment induces a decrease in the release of norepinephrine after electrical stimulation of rat brain [153]. Interestingly, this effect can be overcome with Ca^{2+}, suggesting that Li^+ may be in competition with Ca^{2+}. However, it is also possible that Li^+ may be competing with Mg^{2+}, since this ion is required for the storage of norepinephrine in intraneuronal granules and since it is thought that Mg^{2+}-activated ATPases may be involved in the release of norepinephrine. Chronic Li^+ treatment appears to have no significant effect on the turnover of the catecholamines in rat brain. In humans, the acute effect of Li^+ appears to depend upon the mental state of the patient [154]: in manic patients, there is a decrease in urinary dopamine, and norepinephrine and its metabolites, whereas in depressed patients there is an increase in some of the metabolites; there is no significant differences in euthymic bipolar patients. Although the data is limited, it appears that chronic Li^+ treatment in patients results in an initial transient increase in the metabolites of norepinephrine, followed by a gradual reduction with no prolonged effects.

The affect of Li^+ on the metabolism of serotonin (5-hydroxytryptamine, 5-HT) is equivocal. A number of studies consistently find a Li^+-induced increase in the levels of the major metabolite, 5-hydroxyindoleacetic acid (5-HIAA), in rat brain and in human CSF [155], which appears to reflect an increase in the rate of synthesis of 5-HT [156]. Li^+-induced increases in the level of the amino acid precursor, tryptophan, and in the uptake of tryptophan by brain have also been reported [157], implying elevated tryptophan availability during Li^+ treatment. In rat brain, chronic Li^+ decreases the activity of tryptophan hydroxylase, the enzyme which, when activated by a Ca^{2+} and calmodulin-dependent protein kinase, leads to the synthesis of 5-HT [158]. Ca^{2+} increases the strength of binding of tryptophan to the enzyme, whereas Li^+ has the opposite effect [159]. Tryptophan uptake is coupled to 5-HT utilization by a negative feedback mechanism and, therefore, the Li^+-induced inhibition of tryptophan hydroxylase with a resultant decrease in 5-HT utilization could produce the observed increase in tryptophan uptake.

The acute and chronic effects of Li^+ in patients are quite different (reviewed by Goodnick [146]). Initially, Li^+ treatment causes an increase in the level of 5-HT and decrease in 5-HT uptake in platelets, with an increase in the level of 5-HIAA in the CSF. The neuroendocrine responses to the 5-HT precursors, tryptophan and 5-hydroxytryptophan [160], and to flenfluramine [161], a 5-HT releaser, are enhanced by Li^+. Thus the combined effect of acute Li^+ treatment is to increase the efficiency of synaptic 5-HT. However, chronic Li^+ administration results in almost the opposite effect, resulting in responses close to the levels found in healthy controls: there is a decrease in 5HT levels, the uptake in platelets is greater than in patients before Li^+ treatment, and the neuroendocrine response to 5HT is reduced.

The effect of Li^+ upon the synthesis and release of acetylcholine in the brain is equivocal: Li^+ is reported to both inhibit and stimulate the synthesis of acetylcholine (reviewed by Wood et al. [162]). Li^+ appears to have no effect on acetyl cholinesterase, the enzyme which catalyzes the hydrolysis of acetylcholine [163]. It has also been observed that the number of acetylcholine receptors in skeletal muscle is decreased by Li^+ [164]. In the erythrocytes of patients on Li^+, the concentration of choline is at least 10-fold higher than normal and the transport of choline is reduced [165]; the effect of Li^+ on choline transport in other cells is not known. A Li^+-induced inhibition of either choline transport and/or the synthesis of acetylcholine could be responsible for the observed accumulation of choline in erythrocytes. This choline is probably derived from membrane phosphatidylcholine which is reportedly decreased in patients on Li^+ [166].

There are also several reports of Li^+-induced effects upon the endorphins (see ref. 162). More recently it has been shown that Li^+ also enhances the activity of tyrosine aminopeptidase in rat pituitary gland [167]. This could result in changes in the levels of the enkephalins which are primarily degraded by aminopeptidases via cleavage of the tyrosine–glycine amide bond [168].

6.2 Endocrine System

Li^+ has been reported to affect virtually every component of the endocrine system to some extent; however any resulting clinical manifestations are very rare [169]. Although these influences do not appear to be related to its mechanism of action in manic-depression, some are involved in the side effects experienced by Li^+-treated patients. Apart from elevated levels of thyroid stimulating hormone (TSH), Li^+ does not appear to affect the basal levels of hormones significantly; however some hormone responses are reported to be altered by Li^+ treatment of bipolar patients [170]. Neuronal activity stimulates the adrenal medulla to release norepinephrine and epinephrine into the blood and, consequently, the plasma from people with mania and depression shows increased levels of both neurotransmitters [171].

Li^+ has variable effects upon all hormones released by the pituitary, either directly or indirectly, and in some cases the effects are equivocal.

The response of luteinizing hormone (LH) to the hypothalamic LH-releasing hormone (LHRH) is reportedly unaffected by Li^+ in manic-depressives, but is elevated in Li^+-treated healthy controls [170]. In the serum of young men treated with Li^+ for aggression, the level of LH is increased while that of testosterone is normal [172]; however in older patients, the levels of testosterone are decreased with no change in LH [173], indicating an age-related effect of Li^+. Data obtained from animal studies suggest that Li^+ interferes with the secretion, rather than the synthesis, of pituitary LH: Li^+ has no effect on LH levels in the pituitary when administered acutely and chronically, but decreases LH levels in the plasma by chronic treatment [174]. In female mice and hamsters, the pro-oestrus surge of LH is significantly suppressed to the extent that the oestrus cycle is totally disrupted [175]. The data for the effects on testosterone levels in rats is contradictory (see ref. 174): Li^+ is reported to result in decreased levels of testosterone by acute treatment, and in decreased or unchanged levels during chronic treatment. Levels of follicle-stimulating hormone are unaffected in rats but are increased in the plasma of Li^+-treated patients [173].

The response of prolactin (PRL) to the hypothalamic TSH-releasing hormone (TRH) and to insulin hypoglycemia is reduced in long-term Li^+-treated patients, as is the response of growth hormone to insulin hypoglycemia; however the concentration of growth hormone is also reported to be elevated during Li^+ treatment [170].

The observed Li^+-induced stimulation of corticotropin (ACTH) secretion from cells in culture, requiring extracellular Ca^{2+}, involves a corresponding and apparently associated increase in the concentration of Ins(1)P, indicating some interaction with phosphoinositide metabolism [176]. Pretreatment with Li^+ desensitizes the cells, reducing this Li^+-induced stimulation of ACTH secretion. Li^+ initially raises plasma cortisol levels in manic-depressives; however the levels are subsequently reduced with chronic Li^+ treatment in both patients and controls [177]. This effect is probably secondary to the stimulation and subsequent desensitization of ACTH secretion by Li^+, as observed in cultured cells.

Reports of the effects of Li^+ upon the thyroid gland and its associated hormones are the most abundant of those concerned with the endocrine system. Li^+ inhibits thyroid hormone release, leading to reduced levels of circulating hormone, in both psychiatric patients and healthy controls [178]. In consequence of this, a negative feedback mechanism increases the production of pituitary TSH. Li^+ also causes an increase in hypothalamic thyroid-releasing hormone (TRH), probably by inhibiting its release. Most Li^+-treated patients do not develop goiter as might be expected from enhanced levels of TSH. However, those with an underlying thyroid defect are more likely to have related problems, such as goiter and hypothyroidism, which are reversible when Li^+ intake is stopped [179]. The effects of Li^+ appear to be primarily at the level of the thyroid gland, preventing it from responding to endogenous TSH stimulation. Li^+ inhibits the TSH-mediated stimulation of adenylate cyclase activity in animal thyroid membranes, including murine [180] and porcine [181], where the extent of inhibition is inversely proportional to the Mg^{2+} concentration, indicating that Li^+ is probably in competition with Mg^{2+} at the enzyme's binding site.

The concentration of Li^+ in the thyroid is three to four times that in serum [179]. It is thought that Li^+ may be concentrated in the thyroid gland by a mechanism similar to the incorporation of iodide, I^-, resulting in competition between Li^+ and I^-: the levels of intracellular I^- decrease when those of Li^+ increase, and vice versa [182]. Li^+ inhibits both the ability of the gland to accumulate I^- and the release of iodine from the gland. *In vitro*, Li^+ has no effect on thyroid peroxidase, the enzyme that catalyzes the incorporation of I^- into tyrosyl residues leading to thyroidal hormone synthesis, but does increase the activity of iodotyrosine–deiodinase, which catalyzes the reductive deiodination of iodotyrosyls, thus maintaining the levels of intracellular I^- [182]. The increase in iodotyrosine–deiodinase activity is probably a response to the Li^+-induced decrease in the concentration of thyroidal I^-. Li^+ has no effect on the conversion of thyroxine to triiodothyronine. The overall effect of this competition between Li^+ and I^- is, therefore, reduced levels of thyroid hormone in the presence of Li^+.

A therapeutic role for Li^+, taking advantage of this inhibitory effect of Li^+ on the thyroid, has been demonstrated in thyrotoxic patients by rapidly lowering the concentration of circulating thyroid hormone [183]; this may be beneficial in the short term treatment of hyperthyroidism. Additionally it has been proposed that Li^+ might augment the effect of ^{131}I therapy in thyroid carcinoma by delaying the release of ^{131}I from the tumor and thus enhancing its beneficial effects. Although the retention of ^{131}I has been observed, any clinical benefits have not yet been clearly demonstrated [184].

Li^+ also has effects upon the peripheral renin–angiotensin axis which is involved in fluid-electrolyte homeostasis. Renin is released from the kidney into the blood, where it cleaves angiotensin (a Ca^{2+}-mobilizing hormone) from angiotensinogen; angiotensin I (AI) is converted to angiotensin II (AII) by angiotensin-converting enzyme. In animals, the activity of plasma renin increases with Li^+ administration, resulting in increases in the levels of circulating AI and AII; however the data for humans is equivocal [185]. AII can cause hypertension; however in rats, although Li^+ causes an acute increase in blood pressure, chronic Li^+ administration does not have predictable effects (see ref. 186). Indeed in spontaneously hypertensive rats, Li^+ actually blocks the increase in blood pressure. AII also causes the release of the antidiuretic hormone, vasopressin, and while Li^+ decreases the response of the kidney to vasopressin due to the inhibition of vasopressin-induced adenylate cyclase activity [187], it also results in an increase in vasopressin secretion, presumably by increasing AII levels [188]. This may explain why two of the observed side effects of Li^+ therapy, polyuria and polydipsia, are not experienced by all patients.

Acute Li^+ administration results in elevated levels of blood glucose, glucocorticoids, and insulin in both animals and humans [189]. Following an initial increase in blood glucose in rats and rabbits, Li^+ induces a marked hypoglycemia which is caused by an increase in the uptake of glucose into cells. This is probably a consequence of the interaction of Li^+ with the glucose transporter in the cell membrane. In humans there is also a reversible increase in blood glucose, with levels returning to normal after 2 hours. In rat hepatocytes, Li^+ stimulates glycogen synthesis and glycogen phosphorylase, the rate-limiting enzymes in glycogen metabolism [190]. There is an increased level of glycogen in the brain and the diaphragm, and almost total disappearance of glycogen in the liver after the initial increase in blood glucose.

The pineal gland and its hormone, melatonin, are sometimes associated with the affective disorders; depressed patients are supersensitive to light, in terms of the intensity and duration of light required to suppress nocturnal melatonin levels [191]. A possible consequence of this is the abnormal circadian rhythms in these patients. In humans, Li^+ inhibits the suppressant effect of light on plasma melatonin levels resulting in elevated daytime levels; however in rats, Li^+ causes a reduction in the nighttime levels of melatonin (see ref. 192). These differences may be simply due to the fact that rats are nocturnal animals. In rats, melatonin levels are markedly reduced in the retina where Li^+ is relatively concentrated, and the circadian rhythm on melatonin levels is abolished [192]. Melatonin is synthesized in the pineal gland and in the retina, following norepinephrine stimulation of cAMP production which subsequently activates pineal *N*-acetyltransferase (NAT), the rate-limiting enzyme in the production of melatonin. The effect of Li^+ upon melatonin levels is therefore probably due to its inhibition of norepinephrine-stimulated adenylate cyclase activity and thus the suppression of NAT activity. In accordance with this, chronic Li^+ administration to rats results in a significant decrease in the activity of NAT; however this effect is not observed with acute treatment [193].

7. HEMATOPOIESIS

Li^+ has significant effects upon hematopoiesis which, although they are probably not related to its therapeutic effect in the treatment of affective disorders, are being harnessed for other potential medical applications. All of the differentiated types of blood cells are derived from common stem cells known as CFU-S (colony forming unit stem), which are primarily generated in the bone marrow (Figure 8). The differentiated blood cells proliferate in response to hormones, which stimulate their specific precursor stem cells to divide. For instance, the progenitor cell specific for the production of erythrocytes is the CFU-E (colony factor unit erythrocyte) and that for both granulocytes and macrophages is CFU-GM. These precursors are dependent upon specific growth factors for their survival, known as colony-stimulating factors, for example, GM-CSF.

The numerous influences of Li^+ upon various aspects of hematopoiesis, *in vivo* and *in vitro*, in animals and in humans, have been the subject of several recent reviews by Gallicchio [194–196]. Li^+ increases the total number of the undifferentiated pluripotent stem cells (CFU-S) derived from murine bone marrow *in vivo* [197] and *in vitro* [198]; it inhibits erythropoiesis [197], reducing the proliferation of erythroid stem cells [199]; it enhances megakaryocytosis, increasing the number of circulating platelets *in vivo* and *in vitro* [200]; and, more significantly, Li^+ stimulates leucocytosis, increasing the number of neutrophils in particular, of eosinophils and, to a lesser extent, of monocytes [201,202]. The increase in neutrophils is due, in part, to enhanced production of the granulocyte progenitor stem cells CFU-GM in the bone marrow [198,203] and to increased production and/or release of the colony-stimulating factor GM-CSF [204]. Li^+ also influences the microenvironment of bone marrow and spleen by increasing the number of stromal colonies *in vivo* [200]. *In vitro*, these colonies support the growth of CFU-GM and of the platelet progenitor cells, colony-forming unit megakaryocytes (CFU-Meg).

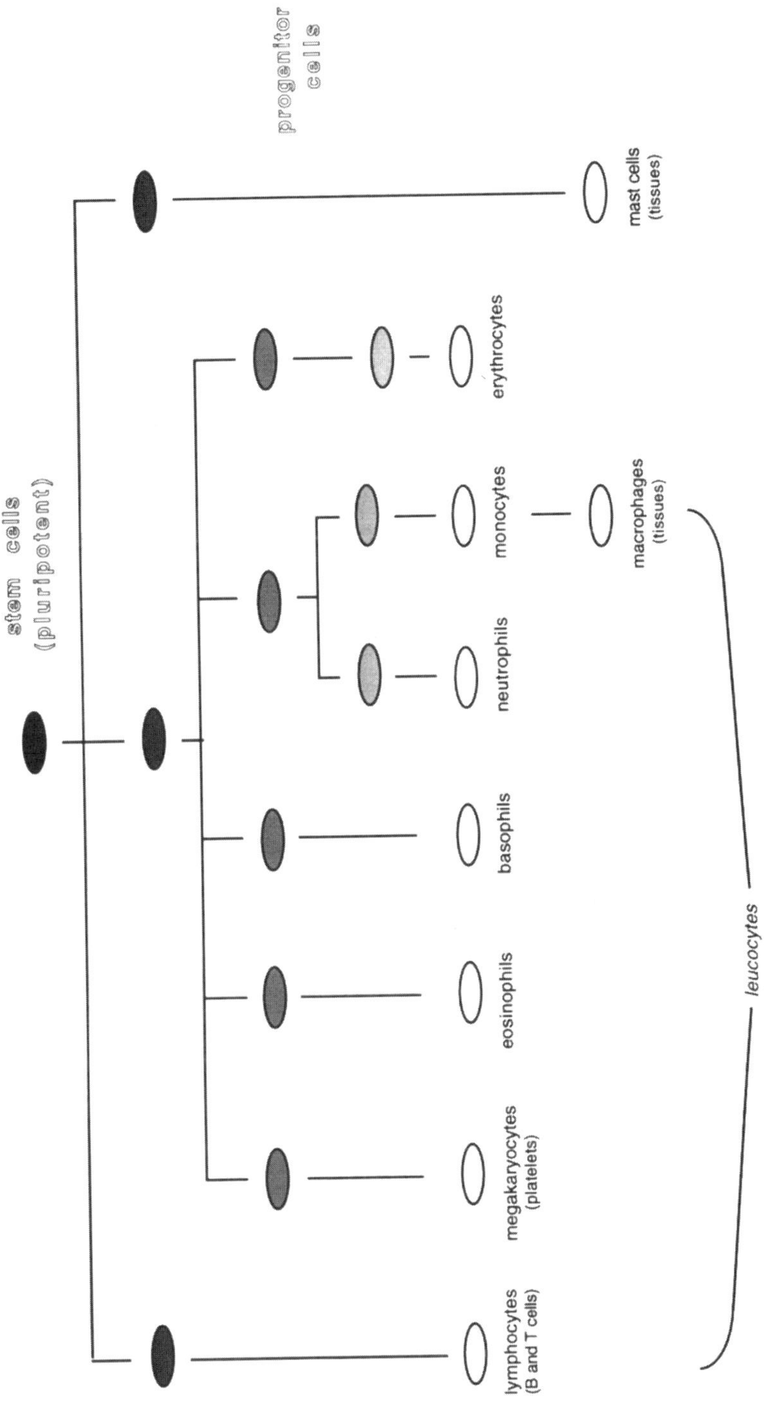

Figure 8. The hematopoietic family tree.

Li^+ also affects the response of lymphocytes to some mitogens. *In vitro*, Li^+ has little effect on lymphocytes in mice, rats, and humans; however, in hamsters the stimulation of lymphocytes is enhanced by the presence of Li^+ [205]. In patients with immunodeficiency due to excess suppresser T-cell activity, Li^+ is effective in inactivating this excess *in vitro* and, subsequently, the number of β-lymphocytes increases; however *in vivo* the immunodeficiency is unaffected [206]. There is no apparent consistent, Li^+-induced effect on the levels of the immunoglobulins.

Recently in a study of long-term Li^+ patients, the levels of neutrophils, helper T-cells, β-lymphocytes, and NK cells were all significantly higher than normal, indicating that these, generally favorable, quantitative changes in leucocyte populations persist with Li^+ therapy [207].

The effects of Li^+ upon hematopoiesis have been proposed to be due to two different systems: modification of the activity of the membrane Na^+/K^+-ATPase, and the inhibition of adenylate cyclase. Monovalent cation flux, in particular Na^+ transport, is known to influence the differentiation and proliferation of hematopoietic stem cells. For instance, ouabain, an effective inhibitor of the membrane Na^+/K^+–ATPase, blocks the proliferation of lymphocytes and has been shown to attenuate the Li^+-induced proliferation of granulocyte precursors [208]. Conversely, Li^+ can reverse the actions of amphotericin and monensin, which mediate Na^+ transport and which inhibit CFU-GM, CFU-E, and CFU-MK colony formation in the absence of Li^+ [209]. Therefore, the influence of Li^+ upon normal physiological cation transport—for example, its influence upon Na^+/K^+–ATPase activity—may be partly responsible for the observed interference in hematopoiesis.

The inhibitory effect of Li^+ upon adenylate cyclase activity may also be put forward to explain both the observations of a decreased rate of erythropoiesis and of enhanced granulopoiesis. It has been reported that therapeutic Li^+ inhibits suppressor cells by decreasing the activity of cAMP [210]. Increased levels of exogenous dibutyryl cyclic AMP (dcAMP), a less polar derivative of cAMP which is thus able to enter cells more easily, is known to both stimulate erythropoiesis and to inhibit granulopoiesis. Therefore Li^+, by inhibiting adenylate cyclase and causing reduced levels of cAMP, might be expected to have the opposite effect to dcAMP, inhibiting erythropoiesis and enhancing granulopoiesis. However, despite the fact that the addition of adenine nucleotides which stimulate granulopoiesis *in vitro* result in enhanced granulopoiesis in the presence of Li^+, the addition of exogenous dcAMP does not significantly block the Li^+-stimulated granulopoiesis [211]. Therefore, the effect of Li^+ upon the hematopoietic system is more complex than the causal decrease in cAMP levels and is probably multifactorial.

The overall effect of Li^+ on the hematopoietic system is of stimulation of the immune system. Not surprisingly then, Li^+ is reported to exacerbate the activity of a number of autoimmune diseases, such as psoriasis [212] and rheumatoid arthritis [213], and to result in the production of autoantibodies in some patients [214]. However, there is no evidence that Li^+'s stimulation of the immune system leads to any reduction in the occurrence of viral or bacterial infections in patients on Li^+ therapy.

In light of the above observations, it appears that Li^+ has a potential role in the treatment of some blood disorders (reviewed in ref. 215). The discovery of the stimulatory effect of Li^+ on granulopoiesis lead to research into its potential use as an adjuvant in reversing abnormal hematopoietic conditions that result in a deficiency of granulocytes. Li^+ can improve both granulocytopenia and leukopenia; however limited success has been achieved in the treatment of aplastic anemia [216]. Prior administration of Li^+ has been used with complete and partial success in hematopoietic recovery following both chemotherapy [217] and radiation therapy [218] in both animal and clinical studies. The potential efficacy of Li^+ in enhancing recovery after bone marrow transplantation is also very promising. Gallicchio has recently shown that pretreatment of donor mice with Li^+ increases hematopoietic recovery in the recipient after the transplant by (1) increasing the number of the CFU-GM and CFU-Meg progenitors in several tissues, and (2) by increasing the number of the circulating neutrophils and platelets [219].

AIDS is associated with aberrant lymphocyte production and it has been proposed that Li^+ may have a potential role in reversing this. Additionally, 3′-azido-3″deoxythymidine (AZT, zidovudine), an effective inhibitor of viral reverse transcriptase that reduces mortality in AIDS patients, induces hematopoietic suppression in patients resulting in anemia, neutropenia, and overall bone-marrow failure [220]. In murine AIDS, the coadministration of Li^+ effectively moderates this toxicity of AZT *in vivo* [221, 222]. There are several case reports where Li^+ has been administered to help reduce the hematopoietic suppression in HIV-infected patients taking AZT (for example, see ref. 223). To date, the use of Li^+ has been limited to a few weeks of treatment, and varying degrees of success have been achieved; nevertheless the outlook in this field is quite hopeful.

8. OTHER ORGANISMS

Nearly all classes of organisms studied exhibit physiological responses to the presence of Li^+ and many of these are detailed in a review by Wissocq et al. including bacteria, viruses, amphibians, fish, birds, fungi, and higher plants [224]. Of particular interest is the inhibitory action of Li^+ upon DNA viruses and its efficacy in the treatment of herpes simplex virus (HSV). A recent review covering the possible mechanisms of action of Li^+ in plants has also been published [225]. Below are merely a few examples to demonstrate the variety of the influences of Li^+.

The presence of Li^+ results in the reversible inhibition of the hormone-induced meiotic maturation of starfish oocytes [226], and also reverses the inhibitory effect of forskolin on the spontaneous resumption of meiosis in uncoated mouse oocytes *in vitro* [227]. In the latter case, the effect of Li^+ upon the adenylate cyclase response occurs only after it has been activated by forskolin.

In bacteria and yeasts, Li^+ has strain-dependent, inhibitory, and morphological effects upon growth. The driving force behind the transport of carbohydrates and amino acids in bacteria is the proton gradient, and in both *E. coli* [228] and *Salmonella typhimurium* cells [229], Li^+ stimulates the movement of proline into cells via a Li^+/proline symport and the transport of melibiose via a cotransport pathway [230]. In both cases, Li^+ is replacing Na^+ and results in the inhibition of growth.

Li^+ causes malformation and affects regeneration in some protozoans, for example *Tetrahymena pyriformis* [231], by inhibiting both DNA and RNA synthesis; by affecting regeneration in *Hydra* [232], planarians [233], and annelids [234]; and by affecting cell signaling mediated by both inositol phosphate metabolism [235] and adenylate cyclase [236] in slime molds.

Lithium succinate ointment is effective against seborrhoeic dermatitis in humans, apparently reducing the rate of sebum production [237]. It was thought that this might be due to the inhibitory effect of Li^+ upon the fungus, *Pityrosporum ovale* (*P. ovale*), whose proliferation is responsible for seborrhoeic dermatitis; however it was found that Li^+ had little effect upon this fungus *in vitro*. It is now believed that the effect of topical Li^+ is to deprive the *P. ovale* of essential nutrients for its growth, in particular the free fatty acids from the skin. Li^+ inhibits the release of all free fatty acids in rat mesenteric artery, and it has also been proposed that Li^+ regulates the conversion of the fatty acid precursors to eicosanoids by suppressing the formation of PGE1 derived from di-GLA and also thromboxane B2 from arachidonic acid [238, 239].

8.1 Viruses

Li^+ has significant inhibitory effects upon DNA viruses, in particular HSV which has been studied in depth. It was originally shown that Li^+ inhibits viral replication in a dose-dependent, reversible manner in HSV-infected baby hamster kidney cells [240], and this has been found to be due to a Li^+-induced decrease in the synthesis of viral DNA [241]. It is now well established that Li^+ inhibits DNA synthesis in HSV types 1 and 2 and in several other DNA viruses, including measles, vaccinia, adenovirus, poxvirus, pseudorabies virus, Epstein-Barr virus, and the bovine, equine, and canine HV's [241]. Interestingly, Li^+ has no effect on the replication of RNA viruses, such as influenza or encephalomyocarditis virus.

Although Li^+ inhibits the synthesis of viral DNA in HSV-infected cells, it has no effect on that of the host cell DNA, making it an ideal drug for this infection. The synthesis of both viral and host cell polypeptides continues in, but is influenced by, the presence of Li^+. In uninfected cells, the synthesis of the host cell polypeptides is unaffected by Li^+. However in HSV-infected cells, the virus itself suppresses protein synthesis in the host and it appears that Li^+ has the ability to reduce this suppression slightly [242]. Li^+ has both stimulatory and inhibitory effects upon the synthesis of the viral polypeptides. For instance, the synthesis of HSV glycoprotein C is significantly decreased by approximately 90% by Li^+ treatment.

The Li^+-induced inhibition of the production of the HSV virus may be related to its actions upon viral DNA polymerase production and activity. Li^+ reduces both the synthesis of DNA polymerase in tissue culture and the activity of DNA polymerase *in vitro*, each by about 50%. It has been proposed that Li^+ reduces the biosynthesis of viral polypeptides and nucleic acids, and hence inhibits viral DNA replication by competition with Mg^{2+}, a cofactor of many enzymes [243]. However, the inhibitory effect of Li^+ on HSV replication in tissue culture is not affected by Mg^{2+} levels. A more likely hypothesis is the alteration of the intracellular K^+ levels, possibly modifying levels of the high-energy phosphate compounds by replacement of either Na^+ or K^+ in Na^+/K^+-ATPase [244]. In tissue culture, HSV replication has been shown to be affected by the concentration of K^+, being inhibited by K^+ starvation. Also, the inhibitory effect of Li^+ can be suppressed by increasing the concentration of K^+.

Li^+ is currently administered topically for the relief of HSV and, in addition, it has been demonstrated that the recurrence of HSV infection is inhibited in Li^+-treated patients, indicating another potential prophylactic effect of Li^+ [245]. Ointment containing 8% lithium succinate has been shown to reduce the severity and the incidence of recurrent genital HSV infection in man [246]. It has also been proposed that Li^+ might be efficacious in treating HIV-infected patients, although any benefits have not yet been demonstrated [247]. While HIV is a RNA virus and as such might not be predicted to be affected by Li^+, it is a retrovirus and utilizes a DNA intermediate for its replication and it uses a DNA polymerase.

8.2 Plants

Although Li^+ appears to have no nutritional function, plants have the ability to absorb the cation, the uptake being inhibited by Ca^{2+} [248]. The transport of Li^+ in plant cells has received little attention; its movement can be passive or mediated by Na^+ channels [249] and in some reports Li^+ actually blocks ion channels; for instance it inhibits voltage-dependent K^+ channels in Chara cells [250]. Li^+ is unevenly distributed throughout the plant, being most concentrated in the cell walls [251].

Li^+ has species-, and concentration-dependent toxic effects upon both seed germination and plant growth. In *Acer pseudoplatanus* cells, Li^+ inhibits both the extrusion of H^+ and the biosynthesis of proteins [252], probably by displacing K^+ as it accumulates in the cells. At low levels, Li^+ can affect the transduction of a variety of signals in plants, including response to trauma, motility, geotropism, and circumnutation, and in induction of some plants to flower, apparently acting as an antagonist to K^+ in the latter case [253]. The effect of Li^+ upon phosphoinositide metabolism in plants is reviewed by Badot [225]. At high toxic concentrations of Li^+, phosphoinositide metabolism in plants can be affected. However in comparison to animals, the inositol monophosphatases in plant cells are relatively insensitive to the presence of Li^+. For instance, Li^+ has no effect on $InsP_3$ phosphatases but inhibits $InsP_2$ phosphatase in carrot cells [254].

Thus, as in humans, the physiological effects of Li^+ in all organisms is very wide-ranging. Due to its chemical similarities to the endogenous cations—Na^+, K^+, Mg^2, and Ca^{2+}—it is hardly surprising that effects as numerous and diverse as, among others, cell signaling via both phosphoinositide metabolism and adenylate cyclase activity, ion transport, and electrochemical homeostasis are altered to some extent by this small, pervasive cation. While the search for the definitive therapeutic action of Li^+ in the treatment of bipolar affective disorder continues, a wider understanding of an apparently infinite number of biological processes will continue to be achieved.

ACKNOWLEDGMENTS

The author is grateful to Dr I. Batty for reading the text and for some useful advice.

REFERENCES

[1] Barr, R.D. and Clarke, W.B., Lithium, 5 (1994) 173–180.
[2] Romney, E.M., Wallace, A., Kinnear, J. and Alexander, G. V., Commun. Soil Sci. Plant Anal., 8 (1977) 799–802.
[3] Barr, R.D., Clarke, W.B., Clarke, R.M., Venturelli, J., Norman, R.G. and Downing, R.G., J. Lab. Clin. Med., 121 (1993) 614–619.
[4] Cade, J.F. J., Med. J. Austr., 36 (1949) 349–352.
[5] Hanlon, L.W., Romaine, M., Gilroy, F.J. and Dietrick, J.E., J. Am. Med. Assoc., 139 (1949) 688–692.
[6] Schou, M. in Birch, N. J. (ed.), *Lithium and the Cell: Pharmacology and Biochemistry*, Academic Press, London, 1991, pp. 1–6.
[7] Thau, K., Meszaros, K. and Simhandl, C., Pharmacopsychiatry, 24 (1991) 85–88.
[8] Crundwell, J.K., Lithium, 5 (1994) 193–204.
[9] Sapse, A.-M. and Von Rague Schleyer, P. (eds.), *Lithium Chemistry: A Theoretical and Experimental Overview*, John Wiley and Sons, New York, 1995.
[10] Heslop, R.B. and Robinson, P.L. (eds.), *Inorganic Chemistry*, Elsevier, London, 1963 p. 254.
[11] Cotton, F.A. and Wilkinson, G. (eds.), *Advanced Inorganic Chemistry*, John Wiley and Sons, New York, 1980, Chap.7.
[12] Birch, N.J., Biol. Psychiat., 7 (1973) 269–272.
[13] El-Mallakh, R.S., Lithium, 1 (1990) 87–92.
[14] Naylor, G.J., Corrigan, F.M. and Smith, A.H. W., Br. J. Psychiat., (1987) 656–661.
[15] Coppen, A., Shaw, D.M., Malleson, A. and Costain, R., Br. J. Med., 1 (1966) 71–75.
[16] Trautner, E.M., Morris, R., Noack, C.H. and Gershon, S., Med. J. Austr. 2 (1955) 280.
[17] Hokin-Neaverson, M., Spiegel, D.A., Lewis, W.C., Burckhart, W. A. and Jefferson, J. W., Res. Comm. Psychol. Psychiat. Behav., 1 (1976) 391–402.
[18] Cook, F.L., Caruso, T.C., Byrne, M.P., Bowes, C.W., Speck, D.H., Liotta, C.L., Tetrahedron Lett., 46 (1974) 4029–4032.
[19] Lehn, J.M., Struct. Bonding, 16 (1973) 1.

[20] Chapoteau, E., Czech, B.P., Zazulak, W. and Kumar, A., Clin. Chem., 38 (1992) 1654–1657.

[21] Riddell, F.G. andArumagam, S., Biochim.Biophys. Acta, 945 (1988) 65–72.

[22] Thellier, M. and Wissocq, J.C., in Birch, N.J. (ed.), *Lithium and the Cell: Pharmacology and Biochemistry*, Academic Press, London, 1991, pp. 59–84.

[23] Hughes, M.S. and Birch N.J., Acad. Sci. Paris Ser III, 314 (1992) 153–158.

[24] Riddell, F.G., Patel, A. and Hughes, M.S., J. Inorg. Biochem., 39 (1990) 187–192.

[25] Sherman, W.R., Munsell, L.Y. and Wong, Y.H., J. Neurochem., 42 (1984) 880–882.

[26] Blijenberg, B.G. and Leijnse, B., Clin. Chim. Acta., 19 (1968) 97–99.

[27] Birch, N.J., Freeman, M.S., Phillips, J.D. and Davie, R.J., Lithium, 3 (1992) 133–137.

[28] Xie, R.Y. and Christian, G.D., Anal. Chem., 58 (1986) 1806–1810.

[29] Thomas, R.C., Simon, W. and Oehme, M., Nature, 258 (1975) 754–756.

[30] Espanol, M.C. and Mota de Freitas, D., Inorg. Chem., 26 (1987) 4356–4359.

[31] Bramham, J., Carter, A.N. and Riddell, F.G., J. Inorg. Biochem., 61 (1996) 273–284.

[32] Thomas, G.M.H., Hughes, M.S., Partridge, S., Olufunwa, R.I., Marr, G. and Birch, N.J., Biochem. Soc. Trans., 16 (1988) 208.

[33] Murakami, M., Seo, Y., Matsumoto, T., Ichikawa, O., Ikeda, A. and Watari, H., Jpn. J. Physiol., 36 (1986) 1267–1274.

[34] Bramham, J., Ph.D Thesis (1993), University of St Andrews.

[35] Preece, N.E., Gadian, D.G., Houseman, J. and Williams, S.R., Lithium, 3 (1992) 287–297.

[36] Ramaprasad, S., Newton, J.E.O., Cardwell, D., Fowler, A.H. and Komoroski, R. A., Mag. Res. Med., 25 (1992) 308–318.

[37] Renshaw, P.F. and Wicklund, S., Biol. Psychiat., 23 (1988) 465–475.

[38] Komoroski, R.A., Newton, J.E.O., Walker, E., Cardwell, D., Jagannathan, N.R., Ramaprasad, S. and Sprigg, J., Mag. Res. Med., 15 (1990) 347–356.

[39] Gyulai, L., Wicklund, S. W., Greenstein, R., Bauer, M.S., Ciccione, P., Whybrow, P.C., Zimmerman, J., Kovachich, G. and Alves, W., Biol. Psychiat., 29 (1991) 1161–1170.

[40] Kato, T., Takahashi, S. and Inubushi, T., Psychiat. Res. Neuroimag., 45, (1992) 53–63.

[41] Renshaw, P.F., Haselgrove, J.C., Leigh, J.S. and Chance, B., Mag. Res. Med., 2 (1985) 512–516.

[42] Schou, M., Ann. Rev. Pharmacol. Toxicol., 16 (1976) 231–253.

[43] Birch, N.J., Clin. Sci. Mol. Med., 46 (1974) 409–413.

[44] Berens, S.C. and Wolff, J., in Johnson, F.N. (ed.), *Lithium Research and Therapy*, Academic Press, London, 1975, Chap. 27.

[45] Birch, N.J. and Hullin, R.P., Life Sci., 11 (1972) 1095–1099.

[46] Maggs, R., Br. J. Psychiat., 109 (1963) 56–65.

[47] Spirtes, M.A., Pharmacol. Biochem. Behavior, 5 (1976) 143–147.

[48] Francis, R.I. and Trail, M.A., Lancet, (1970) 523–524.

[49] Thellier, M., Wissocq, J.C. and Heurteaux, C., Nature, 283 (1980) 299–302.

[50] Schou, M., Acta Pharmacol. Toxicol., 15 (1958) 115–124.

[51] Diamond, J.M., Ehrlich, B.E., Morawski, S.G., Santa Ana, C.A. and Fordtran, J. S., J. Membr. Biol., 72 (1983) 153–159.
[52] Phillips, J.D., Davie, R.J., Kmiot, W.A., Poxon, V.A., Keighley, M.R.B. and Birch, N.J., Br. J. Pharmacol., 96 (1988) 253P.
[53] Rybakowski, J.K., Chlopicka, M., Kapelski, Z., Hernacka, B., Szajnerman, Z. and Kasprzak, K., Int. Pharmacopsychiatr., 9 (1974) 166–171.
[54] Gani, D., Downes, C.P., Batty, I. and Bramham, J., Biochem. Biophys. Acta, 1177 (1993) 253–269.
[55] Duhm, J. and Becker, B.F. in Brewer, G. (ed.), *The Red Cell*, Alan R. Liss, New York, 1978, pp. 551–570.
[56] Duhm, J. and Becker, B.F., Pflugers Arch., 367 (1977) 211–219.
[57] Funder, J., Tosteson, D.C. and Wieth J.O., J. Gen. Physiol., 80 (1978) 721–746.
[58] Haas, M., Schooler, J. and Tosteson, D.C., Nature, 258 (1975) 425–427.
[59] Duhm, J., Eisenried, F., Becker, B.F. and Greil, W., Pflugers Arch., 364 (1976) 147–155.
[60] Sarkadi, B., Allimoff, J.K., Gunn, R.B. and Tosteson, D.C., J. Gen. Physiol., 72 (1978) 249–265.
[61] Canessa, M., Bise, I., Adragna, N. and Tosteson, D.C., J. Gen. Physiol., 80 (1982) 149–168.
[62] Rybakowski, J.K., Lithium, 1 (1990) 75–85.
[63] Rybakowski, J.K., Frazer, A., Mendels, J. and Ramsey, T.A., Commun. Psychopharmacol., 2 (1978) 99–104.
[64] Rybakowski, J.K. and Strzyzewski, W., Lancet, i (1976) 1408–1409.
[65] Elizur, A., Shopsin, B., Gershon, S. and Ehlenberger, A., Clin. Pharmacol. Ther., 13 (1972) 947–953.
[66] Dorus, E., Pandey, G.N. and Davies, J.M., Arch. Gen. Psychiat., 32 (1975) 1097–1102.
[67] Rybakowski, J.K. and Lehmann, W., Biol. Psychiat., 29 (1991) 340–346.
[68] Pandey, G.N., Dorus, E., Casper, R.C., Janicek, P. and Davies, J.M., Prog. Neuropsychopharmacol. Biol. Psychiat., 8 (1984) 547–555.
[69] Ehrlich, B.E., Diamond, J.M. and Gosenfeld, L., Biochem. Pharmacol., 30 (1981) 2539–2543.
[70] Duhm, J. and Becker, B.F., J. Memb. Biol., 51 (1979) 263–286.
[71] Gorkin, R.A. and Richelson, E., Neuropharmacol., 20 (1981) 791–801.
[72] Reiser, G., Scholz, F. and Hamprecht, B., J. Neurochem., 39 (1982) 228–234.
[73] Gorkin, R.A. and Richelson, E., Brain Res., 171 (1979) 365–368.
[74] Reiser, G., and Duhm, J., Brain Res., 252 (1982) 247–258.
[75] Szentistvanyi, I., Janka, Z., Rimanoczy, A., Latzkovits, L. and Juhasz, A., Cell Mol. Biol., 25 (1980) 315–321.
[76] Janka, Z., Szentistvanyi, I., Rimanoczy, A. and Juhasz, A., Psychopharmacology, 71 (1980) 159–163.
[77] Saneto, R.P. and Perez-Polo, J. R., J. Neurosci. Res., 7 (1982) 413–418.
[78] Berridge, M.J., Downes, C.P. and Hanley, M.R., Biochem. J., 206 (1982) 587–595.
[79] Berridge, M.J., Ann. Rev. Biochem., 56 (1987) 159–193.
[80] Drummond, A.H., Trends Pharmacol. Sci., 8 (1987) 129–133.

[81] Ragan, C.I., in Bach, R.O. and Gallichio, V.S. (eds.), *Lithium and Cell Physiology,* Springer-Verlag, New York, 1990, pp. 102–120.
[82] Whitworth, P. and Kendall, D.A., J. Neurochem., 51 (1988) 258–265.
[83] Shears, S.B., Biochem. J., 260 (1989) 313–324.
[84] Downes, C.P. and Carter, A.N., Cell Signal., 3 (1991) 501–513.
[85] Sherman, W.R., in Birch, N. J. (ed.), *Lithium and the Cell: Pharmacology and Biochemistry*, Academic Press, London, 1991, pp. 121–157.
[86] Manji, H.K., Potter, W.Z. and Lenox, R.H., Arch. Gen. Psychiat., 52 (1995) 531–543.
[87] Batty, I. and Nahorski, S.R., J. Neurochem., 45 (1985) 1514–1521.
[88] Batty, I. and Nahorski, S.R., Biochem. J., 247 (1987) 797–800.
[89] Allison, J.H. and Stewart, M.A., Nature New Biol., 233 (1971) 267–268.
[90] Allison, J.H., Blisner, M.E., Holland, W.H., Hipps, P.P. and Sherman, W.R., Biochem. Bipohys. Res. Commun., 71 (1976) 664–670.
[91] Hallcher, L.M. and Sherman W.R., J. Biol. Chem., 255 (1980) 10896–10901.
[92] Nahorski, S.R., Ragan, C.I. and Challiss, R.A. J., Trends Pharmacol. Sci., 12 (1991) 297–303.
[93] Kennedy, E.D., Challiss, R.A.J., Ragan, C.I. and Nahorski, S.R., Biochem. J., 267 (1990) 781–786.
[94] Jope, R.S., Song, L. and Kolasa, K., Biol. Psychiat., 31 (1992) 505–514.
[95] Downes, C.P. and Stone, M.A., Biochem. J., 236 (1986) 895–901.
[96] Manji, H.K. and Lenox, R.H., Synapse, 16 (1994) 11–28.
[97] Anderson, S.M.P., Godfrey, P.P. and Grahame-Smith D.G., Br. J. Pharmacol., 93 (1988) 96P.
[98] Wang, H.Y. and Friedman, E., Psycopharmacol., 99 (1989) 213–218.
[99] Kendall, D.A. and Nahorski, S.R., J. Pharmacol. Exp. Ther., 241 (1987) 1023–1027.
[100] Sherman, W.R. in Michell, R.H., Drummond, A.H. and Downes, C.P. (eds.), *Inositol Lipids and Cell Signalling*, Academic Press, London, 1989, pp. 39–79.
[101] Berridge, M.J., J. Amer. Med. Assoc., 262 (1989) 1834–1841.
[102] Nishizuka, Y., Science, 258 (1992) 607–614.
[103] Agam, G., Shapiro, Y., Bersudsky, Y., Kofman, O. and Belmaker, R. H., Pharmacol. Biochem. Behav., 49 (1994) 341–343.
[104] Berridge, M.J., Downes, C.P. and Hanley, M.R., Cell, 59 (1989) 411–419.
[105] Leech, A.P., Baker, G.R., Shute, J.K., Cohen, M.A. and Gani, D., Eur. J. Biochem., 212 (1993) 693–704.
[106] Takimoto, K., Okada, M., Matsuda, Y. and Nakagawa, H., J. Biochem.(Tokyo), 98 (1985) 363–370.
[107] Gee, N.S., Ragan, C.I., Watling, K.J., Aspley, S., Jackson, R.G., Reid, G.G., Gani, D. and Shute, J.K. Biochem. J., 249 (1988) 883–889.
[108] Shute, J.K., Baker, G.R., Billington, D.C. and Gani, D.J, Chem. Soc. Chem. Commun., (1988) 626–628.
[109] Ganzhorn, A.J. and Chanal, M.-C., Biochemistry, 29 (1990) 6065–6071.
[110] Honchar, M.P., Ackermann, K.E. and Sherman W.R., J. Neurochem., 53 (1989) 590–594.
[111] Cornish-Bowden, A., FEBS Lett., 203 (1986) 3–6.

[112] McAllister, G., Whiting, P., Hammond, E.A., Knowles, M.R., Atack, J.R., Bailey, F.J., Maigetter, R. and Ragan, C.I., Biochem. J., 284 (1992) 749–752.
[113] Diehl, R.E., Whiting, P., Potter, J., Gee, N., Ragan, C.I., Linemeyer, D., Schoepfer, R., Bennett, C. and Dixon, R.A.F., J. Biol. Chem., 94 (1990) 5946–5949.
[114] Bone, R., Springer, J.P. and Atack, J.R., Proc. Natl. Acad. Sci. USA, 89 (1992) 10031–10035.
[115] Cole, A.G. and Gani, D., J. Chem. Soc. Chem. Commun., (1994) 1139–1141.
[116] Strasser, F., Pelton, P.D. and Ganzhorn, A. J., Biochem. J., 307 (1995) 585–593.
[117] Pollack, S.J., Knowles, M.R., Atack, J.R., Broughton, H.B., Ragan, C.I., Osborne, S.A., and McAllister, G., Eur. J. Biochem., 217 (1993) 281–287.
[118] Gore, M.G., Greasley, P. McAllister, G. and Ragan, C.I., Biochem. J., 296 (1993) 811–815.
[119] Bone, R., Frank, L., Springer, J.P., Pollack, S.J., Osborne, S.A., Atack, J. R., Knowles, M., McAllister, G., Ragan, C.I., Broughton, H.B., Baker, R. and Fletcher, S.R., Biochemistry, 33 (1994) 9460–9467.
[120] Bone, R., Frank, L., Springer, J.P. and Atack, J.R., Biochemistry, 33 (1994) 9468–9476.
[121] York, J.D. and Majerus, P.W., Proc. Natl. Acad. Sci. USA, 87 (1990) 9548–9552.
[122] York, J.D., Chen, Z., Ponder, J.W., Chauhan, A.K., Mathews, F.S. and Majerus, P.W., J. Mol. Biol., 236 (1994) 584–589.
[123] York, J.D., Ponder, J.W., Chen, Z., Mathews, F.S. and Majerus, P.W., Biochemistry, 33 (1994) 13164–13171.
[124] Levitzki, A., Cell Biophys., 12 (1988) 133–143.
[125] Limbard, L.E., FASEB J., 2 (1988) 2686–2695.
[126] Birnbaumer, L., Abramowitz, J., Yatani, A., Okabe, K., Mattera, R., Graf, R., Sandford, J., Codina, J. and Brown, A.M., Crit. Rev. Biochem. Mol. Biol., 25 (1990) 225–244.
[127] Sulakhe, P.V. and Hoehn, E.K., Int. J. Biochem., 16 (1984) 1029–1035.
[128] Yoshimasa, T., Sibley, D.R., Bouvier, M., Lefkowitz, R.J. and Caron, M.G., Nature, 327 (1987) 67–70.
[129] Choi, E.J. and Toscano Jr, W.A., J. Biol. Chem., 263 (1988) 17167–17172.
[130] Mork, A. and Geisler, A., Neuropharmacology, 28 (1989) 307–311.
[131] Mork, A., Lithium, 1 (1990) 131–148.
[132] Palmer, G.C., Robinson, G.A., Manian, A.A. and Sulser, F., Psychopharmacologia, 23 (1972) 201–211.
[133] Ebstein, R.P., Hermoni, M. and Belmaker, R.H., J. Pharmacol. Exp. Ther., 213 (1980) 161–167.
[134] Newman, M., Klein, E., Birmaher, B., Feinsod, M. and Belmaker, R.H., Brain Res., 278 (1983) 380–381.
[135] Stefanini, E., Longoni, R., Fadda, F., Spano, P.F. and Gessa, G.L., J. Neurochem., 30 (1978) 257–258.
[136] Singer, I., Rotenberg, D. and Puschett, J.B., J. Clin. Invest., 51 (1972) 1081–1091.
[137] Avissar, S., Schreiber, G., Dennon, A. and Belmaker, R.H., Nature, 31 (1988) 440–442.
[138] Dousa, T. and Hechter, O., Lancet, 1 (1970) 834–835.
[139] Mork, A. and Geisler, A., Lithium, 2 (1991) 127–133.

[140] Cech, S.Y. and Maguire, M.E., Mol. Pharmacol., 22 (1982) 267–272.

[141] Mork, A. and Geisler, A., Pharmacol. Toxicol. 60 (1987) 241–248.

[142] Levell, M.J. and Hullin, R.P., in Birch, N.J. (ed.), *Lithium and the Cell: Pharmacology and Biochemistry*, Academic Press, London, 1991, pp. 327–342.

[143] Schildkraut, J.J., Amer. J. Psychiat., 122 (1965) 509–522.

[144] Bunney, W.E., Wehr, T.A., Gillin, J.C., Ann. Intern. Med., 87 (1977) 319–335.

[145] Kennedy, S.H., Tighe, S., McVey, G. and Brown, G.M., J. Nerv. Ment. Dis., 177 (1989) 300–303.

[146] Goodnick, P., Lithium, 1 (1990) 65–73.

[147] Odagaki, Y., Koyama, T. and Yamashita, I., Lithium, 3 (1992) 95–107.

[148] Geisler, A., Mork, A. and Klysner, R., in Birch, N. J. (ed.), *Lithium: Inorganic Pharmacology and Psychiatric Use*, IRL Press, Oxford, 1988, pp. 195–200.

[149] Rosenblatt, J., Pert, A., Layton, B. and Bunney, W.E., Eur. J. Pharmacol., 67 (1980) 321–322.

[150] Linnoila, M., Karoum, F. and Potter, W.Z., Arch. Gen. Psychiat., 40 (1983) 1015–1017.

[151] Colburn, R.W., Goodwin, F.K., Bunney, W.E., Jr., and Davis, J.M., Nature, 215 (1967) 1395–1397.

[152] Perez-Cruet, J., Tagliamonte, A., Tagliamonte, P and Gessa, G., J. Pharmacol. Exp. Therap., 178 (1971) 325.

[153] Katz, R.I. and Kopin, I.J., Biochem. Pharmacol., 18 (1969) 1835–1839.

[154] Greenspan, K., Schildkraut, J.J. and Gordon, E.K., J. Psychiat. Res., 7 (1970) 171–183.

[155] Bauers, M.B. Jr. and Heninger, G.R., Commun. Psychopharmacol., 1 (1977) 135–145.

[156] Nagahiro, S., Takoda, A., Diksic, M., Sourkes, T.L., Missala, K. and Yamamoto, Y.L., J. Cereb. Blood Flow Metab., 10 (1990) 13–21.

[157] Swann, A.C., Heninger, G.R., Roth, R.H. and Maas, J.W., Life Sci., 28 (1981) 347–354.

[158] Knapp, S. and Mandel, A.J., Science, 180 (1973) 645–647.

[159] Knapp, S. and Mandel, A.J., J. Neural Transm., 45 (1979) 1–15.

[160] Goodnick, P.J., Fieve, R.R. and Schlegel, A., Psychopharmacology, 23 (1987) 510–513.

[161] Muhlbauer, H.D. and Muller-Oerlinghausen, B., J. Neural Transm., 61 (1985) 81–94.

[162] Wood, A.J. and Goodwin, G.M., Psychol. Med., 17 (1987) 579–600.

[163] Simpson, L.L., Psychopharmacologia, 38 (1974) 145–150.

[164] Pestronk, A. and Drachman, D.B., Science, 210 (1980) 342–343.

[165] Jope, R.S., Jenden, D.J., Ehrlich, B.E. and Diamond, J.M., N. Eng. J. Med., 299 (1978) 833–834.

[166] Miller, B.L., Lithium, 1 (1990) 126–128.

[167] De Gandarias, J.M., Irazusta, J., Varora, A., Fernandez, D., Gil, J. and Casis, L., Lithium, 5 (1994) 235–239.

[168] Thorsett, E.D. and Wivratt, M., in Turner, A. J. (ed.), *Neuropeptides and their Peptidases*, Ellis Horwood, Chichester, 1987, pp. 229–292.

[169] Johnson, F.N. (ed.), *Lithium and the Endocrine System*, Karger, Basel, 1988.

[170] Grof, E., Grof, P. and Brown, G.M., in Birch, N. J. (ed.), *Lithium: Inorganic Pharmacology and Psychiatric Use*, IRL Press, Oxford, 1988, pp. 177–181.
[171] Axelrod, J, and Reisine, T.D., Science, 224 (1984) 452–459.
[172] Sheard, M.H., Marini, J.L. and Giddings, S.S., Dis. Nerv. Syst., 38 (1977) 765–7769.
[173] Sanchez, R.S., Murthy, G.G., Mehta, J., Shreeve, W.W. and Singh, F. R., Fert. Steril., 27 (1976) 667–669.
[174] Banerji, T.K., in Johnson, F. N. (ed.), *Lithium and the Endocrine System*, Karger, Basel, 1988, Chap. 5.
[175] Banerji, T.K., Collins, T.J., Parkening, T.A., Rassoli, A.H., Sheika, S. H. and LeGate, L.S., IRCS Med. Sci., 14 (1986) 523–524.
[176] Zatz, M. and Reisine, T., Proc. Natl. Acad. Sci. USA, 82 (1985) 1286–1290.
[177] Smigen, L and Perris, C., Neuropsychobiology, 11 (1984) 219–223.
[178] Perrild, H., Feldt-Rasmussen, U., Hegedus, L. and Baastrup, P.C., in Birch, N.J. (ed.), *Lithium: Inorganic Pharmacology and Psychiatric Use*, IRL Press., Oxford, 1988, pp. 183–187.
[179] Schou, M., Amidsen, A., Jensen, S.E. and Olsen, T., Br. Med. J., 3 (1968) 710–713.
[180] Mori, M., Tajima, K., Ode, Y., Matsui, I., Mashita, K. and Tarui, S., Endocrinology, 124 (1989) 1365–1369.
[181] Urabe, M., Hershman, J.M., Pang, X.P., Murakami, S. and Sugawara, M., Endocrinology, 129 (1991) 807–814.
[182] Teraoka, K., Lithium, 2 (1991) 37–42.
[183] Temple, R., Berman, H., Robbins, J. and Wolff, J., J. Clin. Invest., 51 (1972) 2746–2756.
[184] Gershengorn, M., Izumi, M. and Robbins, J., J. Clin. Endocr. Metab., 42 (1976) 105–111.
[185] Mailman, R.B., Psychopharmacology, 80 (1983) 143–149.
[186] O'Connor, E.F., in Johnson, F. N. (ed.), *Lithium and the Endocrine System*, Karger, Basel, 1988, Chap. 15.
[187] Dousa, T. and Hechter, O., Life Sci., 9 (1970) 765–770.
[188] Mailman, R.B., in Johnson, F. N. (ed.), *Lithium and the Endocrine System*, Karger, Basel, 1988, Chap. 2.
[189] Storlien, L.H., Higson, F.M., Gleeson, R.M., Smythe, G.A. and Atrens, D.M., Pharmacol. Biochem. Behav., 22 (1985) 119–125.
[190] Bosch, F., Gomez-Foix, A.M., Arino, J. and Guinovart, J.J., J. Biol. Chem. 261, 16927–16931.
[191] Lewy, A.J., Nurnberger, J., Wehr, T., Pack, D., Becker, L., Powell, R. and Newsome, D., Am. J. Psychiat., 142 (1985) 725–727.
[192] Seggie, J., in Johnson, F.N. (ed.), *Lithium and the Endocrine System*, Karger, Basel, 1988, Chap. 4.
[193] Yocca, F.D. and Friedman, E., in Johnson, F.N. (ed.), *Lithium and the Endocrine System*, Karger, Basel, 1988, Chap. 18.
[194] Gallichio, V. S., in Birch, N.J. (ed.), *Lithium: Inorganic Pharmacology and Psychiatric Use*, IRL Press, Oxford, 1988, pp. 93–95.
[195] Gallichio, V.S., in Bach R.O. and Gallichio, V.S. (eds.), *Lithium and Cell Physiology*, Springer-Verlag, New York, 1990, pp. 82–93.

[196] Gallichio, V.S., in Gallichio, V. S. (ed.), *Lithium and the Blood*, S. Karger, Basel, 1991, pp. 1–17.
[197] Gallichio, V.S. and Chen, M.G., Blood, 56 (1980) 1150–1152.
[198] Gallichio, V.S. and Chen, M.G. Exper. Hematol., 9 (1981) 804–810.
[199] Gallichio, V.S. and Murphy, M.J., Jr. Cell Tiss. Res., 733 (1983) 175–181.
[200] Gallichio, V.S., Gamba-Vitalo, C., Watts, T.D. and Chen, M. G., J. Lab. Clin. Med., 108 (1986) 199–205.
[201] Rothstein, G., Clarkson, D.R., Larsen, W., Grasser, B. L., Athens, J. W., N. Engl. J. Med., 298 (1978) 178–180.
[202] Perez-Cruet, J., Dancey, J.T. and Warte, J., in Johnson, F. N. and Johnson, S. (eds.), *Lithium in Medical Practice*, MTP Press, Lancaster, 1977.
[203] Hammond, W.P. and Dale, D.C., Blood, 59 (1982) 179–184.
[204] Gallichio, V.S., Chen, M.G. and Watts, T.D., Cell Immunol., 85 (1984) 58–66.
[205] Hart, D.A., in Birch, N.J. (ed.), *Lithium and the Cell: Pharmacology and Biochemistry*, Academic Press, London, 1991, pp. 289–315, and references therein.
[206] Dosch, H.-M., Matheson, D., Shuurman, R. and Gelfand, E., in Rossof, A. and Robinson, W. (eds.), *Lithium Effects on Granulopoiesis and Immune Function*, Plenum Press, New York, 1980 pp. 463–469.
[207] Rybakowski, J.K., Amsterdam, J.D. and Prystrowski, Lithium, 4 (1993) 205–209.
[208] Gallicchio, V.S., Exp. Cell Biol., 52 (1985) 287–293.
[209] Gallicchio, V.S., Lithium, 1 (1990) 93–100.
[210] Gelfand, E.W., Dosch, H.-M., Hastings, B. and Shore, A., Science, 203 (1979) 365–367.
[211] Gualtieri, R.J., Berne, R.M., McGrath, H.E., Huster, W.J. and Quesenberry, P.J., Exp. Hematol., 14 (1986) 689–695.
[212] Horrobin, D.F., in Bach R.O. and Gallichio, V.S. (eds.), *Lithium and Cell Physiology*, Springer-Verlag, New York, 1990, pp. 158–167.
[213] Khanna, R. and Chatterjee, S., J. Clin. Psychiat., 52 (1991) 43–44.
[214] Whalley, L.J., Roberts, D.F., Wentzel, J. and Watson, K. C., J. Affective Disorders, 3 (1981) 123–130.
[215] Ananth, J. and Johnson, K.M., Lithium, 4 (1993) 13–23.
[216] Barrios, N.K., Kirkpatrick, D.V., Stine, K.C. and Humbert, J. R., Br. J. Hematol., 73 (1989) 422–423.
[217] Perez, H.D., Kaplan, H.P., Goldstein, I.M., Shenkman, L. and Borkowsky, W., Clin. Immunopathol., 16 (1980) 308–315.
[218] Lyman, G.H., Williams, C.C., Preston, D., Goldman, A., Dinwoodie, W. R., Saba, H., Hartmann, R., Jensen, R. and Shukovsky, L., Am. J. Med., 70 (1981) 1222–1229.
[219] Gallicchio, V.S., Messino, M.J., Hulette, B.C., Hughes, N. K. and Bieschke, M. K., Lithium, 2 (1991) 27–36.
[220] Mitsuya, H. and Broder, S., Proc. Natl. Acad. Sci. USA, 83 (1986) 1911–1915.
[221] Gallichio, V.S. and Hughes, N.K., J. Int. Med., 231 (1992) 219–226.
[222] Gallichio, V.S., Hughes, N.K. and Tse, K.F., J. Int. Med., 233 (1993) 259–268.
[223] Jordan, W.C., J. Natl. Med. Assoc., 84 (1992) 1044–1046.

[224] Wissocq, J.C., Attais, J. and Thellier, M., in Birch, N.J. (ed.), *Lithium and the Cell: Pharmacology and Biochemistry*, Academic Press, London, 1991, pp. 7–34.
[225] Badot, P.M., Lithium, 3 (1992) 155–168.
[226] Picard, A. and Doree, M., Exp. Cell Res., 147 (1983) 41–50.
[227] Gavin, A.C. and Schorderet-Slatkine, S., Exp. Cell Res., 179 (1988) 298–302.
[228] Kawasaki, T. and Kayama, Y., Biochem. Biophys. Res. Commun., 55 (1973) 52–59.
[229] Stock, J. and Roseman, S., Biochem. Biophys. Res. Commun., 44 (1971) 132–138.
[230] Tsuchiya, t., Lopilato, J. and Wilson, T.H., J. Membr. Biol., 42 (1978) 45–59.
[231] Volm, M., Schwartz, V. and Wayss, K., Naturwissenschaften, 57 (1970) 250.
[232] Hassel, M. and Berking, S.W., Roux's Arch. Dev. Biol., 197 (1988) 471–475.
[233] Bustuoabad. O.D., Fiocchi, M.G. and Matteucci, I.M., Medicina (Buenos Aires), 40 (1980) 547–552.
[234] Stephan-Dubois, F. and Morniroli, C., Bull. Soc. Zool. France, 92 (1967) 335–344.
[235] Peters, D.J.M., Van Lookeren-Campagne, M.M., Van Haastert, P. J. M., Spek, W. and Schaap, P., J. Cell Sci., 93 (1989) 205–210.
[236] Van Lookeren-Campagne, M.M., Wang, M., Spek, W., Peters, D. and Schaap, P., Dev. Gen., 9 (1988) 589–596.
[237] Horrobin, D.F., Lithium, 1 (1990) 149–155.
[238] Horrobin, D.F., Jenkins, D.K., Mitchell, J. and Manku, M.S., in Birch, N. J. (ed.), *Lithium: Inorganic Pharmacology and Psychiatric Use*, IRL Press, Oxford, 1988, pp. 173–176.
[239] Horrobin D.F., in Bach R.O. and Gallichio, V.S. (eds.), *Lithium and Cell Physiology*, Springer-Verlag, New York, 1990, pp. 137–149.
[240] Skinner, G.R.B., Hartley, C.E., Buchan, A., Harper, L., and Gallimore, P., Med. Microbiol. Immunol., 168 (1980) 258–265.
[241] Buchan, A., Randall, S., Hartley, C.E., Skinner, G.R.B. and Fuller, A., in Birch, N.J. (ed.), *Lithium: Inorganic Pharmacology and Psychiatric Use*, IRL Press, Oxford, 1988, pp. 83–90.
[242] Randall, S., Hartley, C.E., Buchan, A., Lancaster, S. and Skinner, G. R. B., in Birch, N. J. (ed.), *Lithium and the Cell: Pharmacology and Biochemistry*, Academic Press, London, 1991, pp. 99–112.
[243] Bach, R.O., Med. Hypoth., 23 (1987) 157–170.
[244] King, L.J., Carl, J.L., Archer, E.G., and Castellanet, M., J. Pharmacol. Exp. Ther., 168 (1969) 163–170.
[245] Rybakowski, J.K. and Amsterdam, J.D., Lithium, 2 (1991) 43–47.
[246] Skinner, G.R.B., Lancet, ii (1983) 288.
[247] Parenti, D.M., Simon, G.L., Scheib, R.G., Proc. 2nd Intern. Symp. AIDS, Paris, (1986).
[248] Jacobsen, L., Moore, D.P. and Hannapel, R.J., Plant Physiol., 35 (1960) 352–358.
[249] Epstein, E. and Hagen, C.E., Plant Physiol., 27 (1952) 457–474.
[250] Tester, M., J. Membr. Biol., 103 (1988) 159–169.
[251] Bielenski, U., Ripoll, C., Demarty, M., Luttge, U. and Thellier, M., Physiol. Plant, 62 (1984) 32–38.

[252] Thoiron, B., Thoiron, A., Luttge, U. and Thellier, M., Biol. Cell., 45 (1982) 39.
[253] Kandeler, R., Planta, 90 (1970) 203–207.
[254] Memon, A.R., Rincon, M. and Boss, W.F., Plant Physiol., 91 (1989) 477–480.

CERULOPLASMIN:
THE BEGINNING OF THE END OF AN ENIGMA

Peter Lindley, Graeme Card,
Irina Zaitseva, and Vjacheslav Zaitsev

Perspectives on Bioinorganic Chemistry
Volume 4, pages 51–89.

ISBN: 0-7623-0352-2

1. INTRODUCTION

Only comparatively recently has it been realized that copper is a vital requirement for the growth, development, and function of most living organisms and that copper and iron metabolisms are intricately linked. An average adult of weight 70 kg has some 110 mg of copper [1] with relatively high concentrations in bone (46 mg) and muscle (26 mg) and smaller concentrations in other organs such as the kidney (3 mg), liver (10 mg), brain (9 mg), and heart (1.5 mg). The blood contains around 6 mg and approximately 95% of this copper resides in a plasma protein, human ceruloplasmin (hCP). Ceruloplasmin, the "sky-blue plasma protein," was first isolated in 1944 [2]. It is a monomeric glycoprotein comprising a single polypeptide chain of 1046 amino acid residues, (molecular weight ≈ 132 kDa) and a glycan content of some 7–8% by weight. The protein belongs to the family of multi-copper oxidases which includes laccase, ascorbate oxidase, and nitrite reductase, and is also related to the blood clotting factors V and VIII. Ceruloplasmin is synthesized in hepatocytes and in normal circumstances six atoms of copper are incorporated before secretion of the protein from the liver into the serum. Defects in hepatic biosynthesis of hCP can result in diseases such as Wilson's disease (hepatolenticular degeneration), which results, if untreated, in copper deposition in the body tissues, particularly the liver, brain, and kidney.

The size and complexity of the hCP molecule are consistent with a multifunctional role for this protein in the plasma, although its precise functions, as yet, remain unclear; for this reason it has been termed the "enigmatic copper protein" [3]. Most of the functions that have been ascribed focus on the presence of the copper centers (see for example refs. 4 and 5), and the major roles can be summarized as follows:

1. ferroxidase activity to mobilize iron for transport via the plasma protein, transferrin, and to eliminate free iron from plasma, thereby protecting the blood and membrane lipids from peroxidative damage,
2. antioxidant activity to remove oxygen-based and other free radicals from the plasma,
3. amine oxidase activity to control levels of biogenic amines in the plasma, cerebral, spinal, and interstitial fluids, and
4. copper transport to deliver the metal to extrahepatic tissues.

It has also been proposed that because of sequence and possible structural homology with the blood clotting factors V and VIII [6] that ceruloplasmin may also participate in blood clotting and its regulation [5].

Extensive biochemical and spectroscopic studies have been undertaken on hCP in order to investigate the nature of the copper centers and their role in structure–function relationships. However, the protein is very susceptible to aggregation, proteolysis, loss of copper, and other chemical degradations and requires careful preparation and handling; in these circumstances it is difficult to review all the literature objectively and comprehensively. A three-dimensional crystal structure of hCP has been reported at a nominal resolution of 3.1 Å [7], but this resolution has been extended to just beyond 3.0 Å. This chapter will summarize some of the more important biochemical and spectroscopic studies of the protein. It will then focus on the structural results recently obtained by X-ray crystallographic methods and attempt to explain putative functions of the protein in terms of its molecular structure.

2. BACKGROUND

2.1 Primary Sequence and Structure Prediction

Human ceruloplasmin can be readily cleaved into three fragments: an N-terminal 67 kDa fragment, a central 50 kDa fragment, and a 19 kDa C-terminal fragment. Completion of the primary structures of the 50 kDa and 19 kDa fragments of hCP enabled Dwulet and Putnam [8] to identify considerable internal homology in the molecule and also to note significant homology with other copper-containing proteins. This led to the proposal that hCP evolved through the fusion of two genes, coding for proteins of around 190 and 160 amino acid residues, respectively. This precursor gene, coding for 350 residues, then triplicated to form the gene for the present-day hCP molecule of around 1050 residues. The determination of the primary sequence of the N-terminal 67 kDa fragment [9] strongly supported an internal triplication of the primary structure of the entire molecule. The polypeptide chain was shown to be divided into three covalently linked homologous segments, each comprising some 340 residues. A sequence identity of some 30% was established between all three segments, but for each pair the identity was closer to 40%. Following the determination of the structure of ascorbate oxidase, AO [10], Messerschmidt and Huber [11] were able to align the sequences of the related blue copper oxidases, laccase, and hCP, on a structural basis. The alignment suggested a three-domain structure for laccase and a six-domain structure for hCP. Furthermore for hCP, potential mononuclear copper binding residues and sites were located in domains 2, 4, and 6, together with a site for a trinuclear cluster at the interface of domains 1 and 6. The relative disposition of the trinuclear cluster and the mononuclear copper in domain 6 was envisaged to be similar to that found in ascorbate oxidase, strongly supporting an oxidase role for hCP. These proposals were developed further by Fenderson et al. [12] based on the structure of nitrite reductase (NR) to give a realistic prediction for the domain structure of hCP.

2.2 Spectroscopic Data

Various spectroscopic methods have been used to probe the nature of the copper centers in the members of the blue copper oxidase family of proteins (e.g. see ref. 13). Prior to the X-ray determination of the structure of ascorbate oxidase in 1989, similarities in the EPR and UV-vis absorption spectra for the blue multi-copper oxidases including laccase and ceruloplasmin had been observed [14] and a number of general conclusions made for the copper centers in ceruloplasmin as shown in Table 1 [13,15]. It was known that six copper atoms were nondialyzable and not available to chelation directly by dithiocarbamate and these coppers were assumed to be tightly bound and/or buried in the protein. Two of the coppers have absorbance maxima around 610 nm and these were interpreted as blue type I coppers with cysteine and histidine ligands, and responsible for the pronounced color of the protein. However, they are not equivalent and one of them, thought to be involved in enzymatic activity, is reduced and reoxidized at a faster rate than the second (e.g. see ref. 16). There was general concurrence that there are two type III

Table 1. Summary of Nature of Copper Sites in hCP from Spectroscopic Data[a,b] Linder [13], Modified from Frieden [15]

No. of Cu Atoms Per Molecule	*Terminology*	*UV-Visible, EPR Detectability*	*Other Properties*
1 × Fast type I	Blue Cu(II), Cu_A	A_{max} at 610 nm EPR detectable	Fast reoxidation, involved in enzyme activity
1 × Slow type I	Blue Cu(II) Cu_B	A_{max} at 610 nm EPR detectable	Slow reoxidation, not involved in enzyme activity.
1 × Type IV	Cu(I or II)	EPR silent	Required by total copper content.
1 × Type II	Permanent Cu(II)	EPR detectable	Binds anions; involved in enzymatic activity.
2 × Type III	Cu(II)	A_{max} at 330 nm EPR silent.	Believed to be spin coupled, in analogy with other multi-copper oxidases
[Additional	Cu(II)	EPR silent	Extraneously bound and readily removable].

Notes: [a]Linder [13].
[b]Modified from Freiden [15].

coppers in ceruloplasmin, assumed to be spin-paired and therefore EPR silent. These coppers are thought to be responsible for the presence of a 330 nm shoulder in the UV-vis absorption spectrum of the protein.

The protein was also thought to contain a type II copper, detectable in low-field EPR measurements and involved in anion binding (*vide infra*). In hCP this copper had been detected in isolated preparations of the protein, but there was controversy regarding its presence *in vivo* in the native protein [17]. The existence of a sixth copper was needed to meet the stoichiometric requirement for six nondialyzable coppers; this was designated type IV and is EPR silent. In addition there was some evidence to suggest extraneous copper which could be removed with Chelex 100 and was thought to be a contaminant acquired during protein isolation [18]. The elucidation of the structure of ascorbate oxidase in 1989 [10] and the spectral similarities in this family of proteins suggested the presence of a four-copper oxidase center in ceruloplasmin. In more recent studies it has been shown that electronic absorption spectroscopy used in conjunction with ligand field theory is one of the most useful general spectroscopic tools for examining the nature of the metal centers in the blue copper oxidases [19].

2.3 Anion Binding

Curzon and Cummings [20] have identified seven categories of inhibitors of ceruloplasmin: inorganic anions, carboxylate anions, -SH compounds, chelating agents, hydrazins, 5-hydroxyindoles, and a miscellaneous group including metal cations. Probably the most informative group is that of the inorganic anions which includes two of the strongest inhibitors of oxidase activity, cyanide and azide, both with inhibitory constants, $K_i \approx 2 \times 10^{-6}$). Azide has been frequently used in attempts to distinguish between ceruloplasmin catalysis of Fe(II) oxidation and other ferroxidase activity in plasma.

The binding of azide and fluoride to the resting enzyme appears to change the EPR signal from the type II copper [21,22]. These results together with binding and kinetic studies [23,24] indicate that one anion binds at low concentration, but two at high concentrations. However, recent X-ray studies on the azide complex with ascorbate oxide [25] clearly indicate that the two azide anions bind to one of the two type III copper atoms. One interpretation of these results could be that during catalysis the trinuclear center acts as a unit and there is little distinction between the type II and III coppers. The binding of azide, cyanate, thiocyanate, and cyanide to ceruloplasmin is associated with the formation of a new absorption band in the region of 375 to 435 nm. At higher concentrations of anion, the blue type I copper 610 nm band decreases to between 50 and 70% of its original value. In contrast, fluoride binding does not cause any shifts in the visible spectrum. Inhibition of oxidase activity by cyanide involves the complete dissociation of copper ions and is not reversible [26]. Calabrese and coworkers [27] have recently shown that the binding of chloride dramatically affects the catalytic efficiency of hCP. At neutral pH, the anion is the activator of the oxidase activity and enhances the catalytic rate by 10-fold. At a pH < 6.0, as previously found, chloride strongly inhibits activity. At intermediate pH values, i.e. around 6.0, the effect is composite with an activating effect at low concentration and an inhibitory effect at higher concentration. Since chloride is present at relatively high concentration in the plasma, these results suggest that hCP in the plasma may be under the control of this anion.

2.4 Copper and Wilson's Disease

Abnormalities in copper metabolism are normally associated with Wilson's disease [28] and Menkes' disease [29,30,31], although total copper and hCP concentrations increase significantly in many inflammatory and infectious diseases including hepatitis and tuberculosis, and a number of different kinds of cancer. A direct connection between copper and coronary artery disease has also been proposed [32]. A useful general review of copper and disease has been given by Linder [33].

Wilson's disease (hepatolenticular degeneration) is an inherited, progressive and, if untreated, ultimately fatal disease resulting in copper deposition in the body tissues, particularly the liver, brain, and kidney. In the disease the level of serum copper is often, but not always, decreased by a factor of about four to five, whereas non-ceruloplasmin bound copper (e.g. albumin) appears to be elevated. Several possibilities have been suggested for the causes of Wilson's disease. These include a decrease in hepatic synthesis of hCP, since some patients appear to possess low serum concentration levels of the protein. Thus, a decrease or retardation in hCP gene expression could lead to reduced hCP synthesis and concomitant reduction in the amount of copper secreted from the liver. However, patients suffering from the disease also exist with normal or near normal levels of ceruloplasmin in the serum so that defects in gene expression are unlikely to be the sole cause. Another possibility involves defects in the mechanisms whereby copper is incorporated into the *apo*-ceruloplasmin. Failure of hCP to incorporate adequate amounts of copper during its synthesis could result in an accumulation of copper in the liver and the release of copper-deficient hCP from this organ could have profound consequences for other organs and enzymes that rely upon hCP as a source of copper. A further explanation for the disease is that there could be defects in the mechanism whereby the copper-loaded hCP is secreted from the liver into the plasma. Copper is also lost from the liver via the bile and this mechanism could be faulty in Wilson's disease. It should be noted that the hCP gene is located on chromosome 8, whereas the putative gene for Wilson's disease appears to be located on chromosome 13 [34]. This would indicate that the basic genetic defect of Wilson's disease does not involve hCP per se, but resides in other genes related to copper metabolism or in the controller gene for ceruloplasmin synthesis. The treatment of the disease is normally aimed at "decoppering" the patient with copper chelators, such as 2,3-dimercaptopropanol and penicillamine, before the excess metal has the opportunity to cause tissue damage.

Ceruloplasmin has long been linked with iron metabolism through its putative role as a ferroxidase [4], but in the last 2 years substantive evidence has evolved to reinforce the links between copper and iron metabolisms; this subject will be pursued in Section 6.

3. FUNCTIONS OF CERULOPLASMIN

It is highly likely that a protein the size of ceruloplasmin has more than one function and a number of functions have been proposed including ferroxidase and antioxidant activity, oxidation of biogenic amines, and copper transport. However, the precise definition of the protein's functions *in vivo* remains elusive. The catalytic cycle of the blue multi-copper oxidases involves a sequential mechanism whereby single electrons are taken from the reducing substrate and transferred via a type I blue copper to the trinuclear center which serves as the dioxygen binding and reduction site. During the reaction cycle the dioxygen is reduced by four electrons to two molecules of water. It is likely that ceruloplasmin undergoes similar reactions whereby cations such as ferrous iron can be oxidized to ferric or their oxidation can be used to reduce oxygen-based free radicals to water. Ceruloplasmin is unique in that it can oxidize both organic and inorganic substrates, and the value for the apparent Michaelis constant varies by six orders of magnitude, from 0.2 μM to 280 mM [35], while the value for V_{max} varies only by one order of magnitude. These observations have prompted the assumption that the limiting step in ceruloplasmin-catalyzed reactions is dependent not on the affinity of the binding site for the substrate, but on the speed of reduction of the "blue" copper in the four-copper oxidase center [36]. The protein is known to catalyze the oxidation of a number of bifunctional aromatic amines and diphenols and recently it has been shown to catalyze the oxidative coupling of 3-hydroxyanthranilic acid to form cinnabarinic acid [37], a phenoxazinone chromophore which protects mammalian tissue from oxidative damage *in vivo*.

3.1 Ferroxidase and Antioxidant Activity

A controversial function of ceruloplasmin *in vivo* is its ability to oxidize ferrous to ferric iron; as a substrate Fe(II) has the lowest apparent K_m and the highest V_{max} of any of the protein's multiple substrates.

Ceruloplasmin has been implicated in mediating the release of ferrous iron from cells prior to its incorporation as ferric iron into the iron transporting protein, serum transferrin [38]. Although this is an attractive proposition, there are a number of problems including the nature of the iron release mechanism from cells, the avidity of transferrin for iron causing auto-oxidation of ferrous iron in the presence of oxygen, and the lack of evidence for a ceruloplasmin–transferrin complex. However, the presence of free ferrous iron (or indeed free divalent copper) in the plasma is to be avoided since it will cause the generation of the highly deleterious OH· radical via a Fenton-type reaction, and the ferroxidase activity of ceruloplasmin, particularly in cells where the oxygen content is low, would prevent such free radical damage. Furthermore, recent studies have identified mutations of the ceruloplasmin gene giving rise to systemic hemosiderosis in humans [39,40], so that the link between ceruloplasmin and iron metabolism and the physiological role of ceruloplasmin as a ferroxidase appears to be firmly established; these matters will be pursued further in Section 6.

Clearly, the ferroxidase activity of ceruloplasmin will lead to the inhibition of free radical formation by inhibiting the interaction of reduced iron species with peroxide, but it is also thought that ceruloplasmin can also react directly with free radical intermediates, thus harmlessly dissipating the extra electrons via the formation of water from dioxygen. Whether ceruloplasmin can react directly with superoxide, rather than indirectly through iron, is not certain, but there is general agreement that the protein is a scavenger of free radicals (e.g. see ref. 41). It is also generally agreed that ceruloplasmin in the serum and other interstitial fluids is important for protection against peroxidative damage and it appears to act as a general neutralizer of reactive oxygen species.

3.2 Copper Transport

The copper transport function of ceruloplasmin has been documented in several reviews (e.g. see refs. 15, 42, 43) and a transport function established. The turnover of ceruloplasmin allows copper ions to move from the major sites of ceruloplasmin synthesis in liver cells [44,45] to peripheral tissues for incorporation into copper-dependent enzymes [46,47], but transport mechanisms may also be active which involve copper atoms in the intact protein. However, the complexity of the protein has made it difficult to determine which, if any, of the six integral copper atoms are involved in copper delivery or whether there exist additional specific transport sites. Solutions and crystals of ceruloplasmin gradually lose their blue color on aging and this could be due either to changes in oxidation state of the blue coppers and/or the depletion of copper from the protein. The subject of additional copper binding sites and copper transport is discussed further in Section 6.

In conjunction with copper transport, specific receptors for ceruloplasmin have been found on blood cells and isolated membranes from a variety of cells [43,48] and a model of copper transport from ceruloplasmin has been constructed (e.g. see ref. 5). In the model, ceruloplasmin binds to a specific membrane receptor. Select copper atoms, not belonging to the four-center ferroxidase site, are reduced to Cu(I) and changes in the protein conformation make them available for binding to groups on the cell membrane. The copper then dissociates from ceruloplasmin by a ligand exchange reaction with membrane sulphydryl sites and is internalized via other sulphydryl groups associated with the integral membrane protein; the ceruloplasmin is released from the membrane surface. The viability of this model has yet to be proved.

4. THE X-RAY STRUCTURE OF HUMAN CERULOPLASMIN, hCP

4.1 Overall Organization of the hCP Molecule

Throughout the description of the structure and the subsequent discussion sections it should be clearly remembered that the resolution of the data is limited to around 3.0 Å and that precise details of the structure cannot be described. A schematic view of the overall organization of the molecule showing the six domains and the distribution of the copper atoms is given in Figure 1a. Figures 1b and 1c show backbone traces of the polypeptide chain viewed along and perpendicular to the pseudo threefold axis of the molecule, respectively. Each of the domains comprises an eight- stranded β-barrel with the strands organized in a manner typical of the small blue cupredoxin proteins such as azurin and plastocyanin [49]; Figure 2 shows the fold for domains 1 and 2, respectively. The odd-numbered domains comprise approximately 200 residues, but the even domains are shorter with only about 150 residues. The barrels are oriented so that they are all the same way up, but the even-numbered domains point inwards towards the pseudo threefold axis reminiscent of the feet of a tripod, whereas the odd-numbered domains point outwards;

this arrangement has a pronounced effect on the separation of the mononuclear copper sites (*vide infra*).

There is considerable similarity between domains 1, 3, and 5 with excellent matching of the β-strands; 168 α-carbon atoms of domain 3 and 149 atoms of domain 5 can be superimposed on domain 1 with a root mean square fit of only 1.0 Å. Major deviations occur in the loops between the first and second, and the fourth and sixth strands. The similarity of the smaller mononuclear copper binding domains, 2, 4, and 6, is even more pronounced with 147 atoms from domain 4 and 143 from domain 6 fitting domain 2 with a root mean square deviation of only 0.9 Å in each case. However, although all 6 domains are based on a cupredoxin-type fold, the various loop regions deteriorate the match between an even and an odd domain with a typical fit of 1.8 Å for only 91 atoms when domain 2 is superposed onto domain 1. The superposition results are summarized as part of Table 2.

The three major cleavage sites in hCP [50] are readily explained in terms of the molecular structure. The 19 kDa fragment corresponds almost exactly to domain 6, so that the cleavage site is interdomain in nature. However, the cleavage site for the larger fragments occurs at residue 479, a residue in domain 3 on an exposed loop prior to strand 6 (see Figure 2a).

4.2 The Copper Binding Sites

The human ceruloplasmin structure contains six integral copper atoms; three of these occupy mononuclear centers in domains 2, 4, and 6, whereas the remaining three form a trinuclear cluster sited at the interface between domains 1 and 6. The trinuclear cluster and the domain 6 mononuclear copper form a four-atom center almost identical to that found in the subunit of ascorbate oxidase [10,51]. In ascorbate oxidase, the organic substrate is subjected to a one-electron oxidation near the mononuclear copper site and in a sequential oxidation of substrate molecules there is a concomitant four-electron reduction of a molecule of oxygen to two molecules of water at the trinuclear center; it is highly likely that hCP catalyzes a similar reaction. Figure 3 shows the distance between the copper atoms, the EPR and UV-vis spectra classification [52], and the binding ligands. All of the copper binding residues are sited either on strand 4 of the β-barrel or on the loop region connecting strands 7 and 8; this is shown in Table 3, together with the corresponding

(a)

(b)

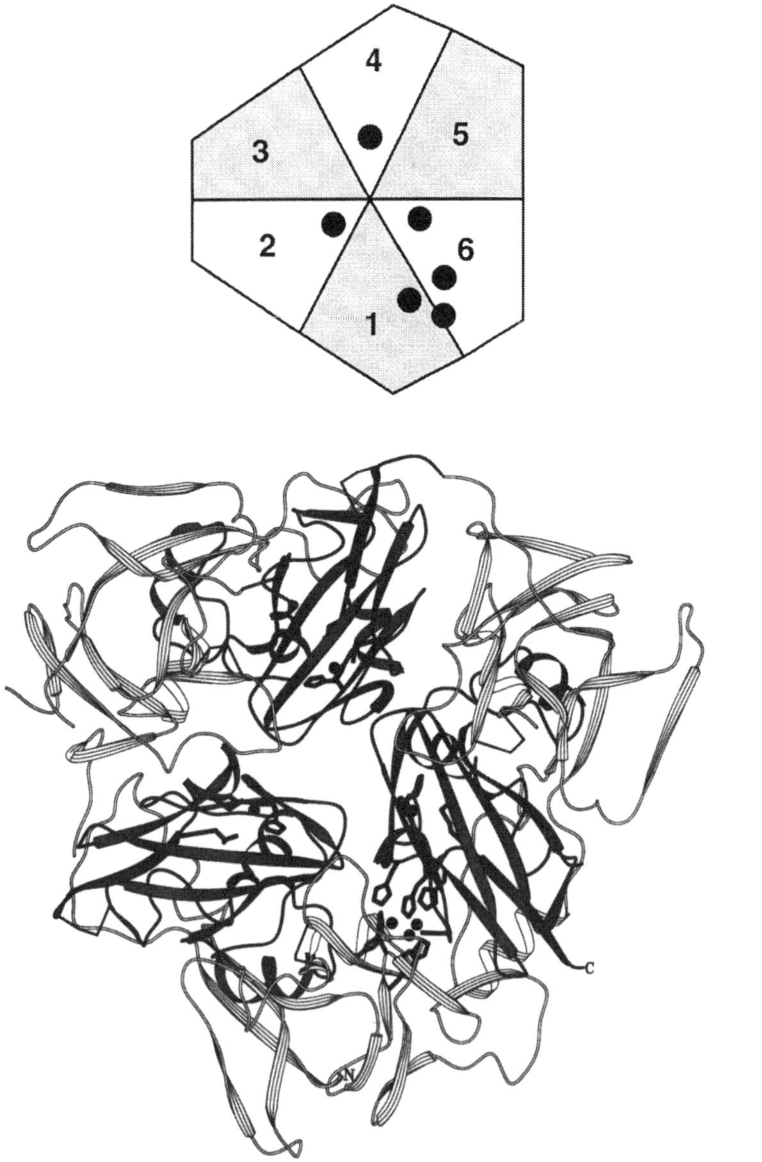

Figure 1. (**a**) A schematic representation of the overall organization of the molecule of human ceruloplasmin. Domains 2, 4, and 6 contain mononuclear copper centers, while the trinuclear copper cluster is located at the interface of domains 1 and 6. (**b**) An α-carbon ribbon diagram of the human ceruloplasmin molecule viewed along the pseudo threefold axis highlighting the triplication of the structure. Domains 1, 3, and 5 are depicted by striped motifs, whereas domains 2, 4, and 6 are dark shaded. The copper

(**c**)

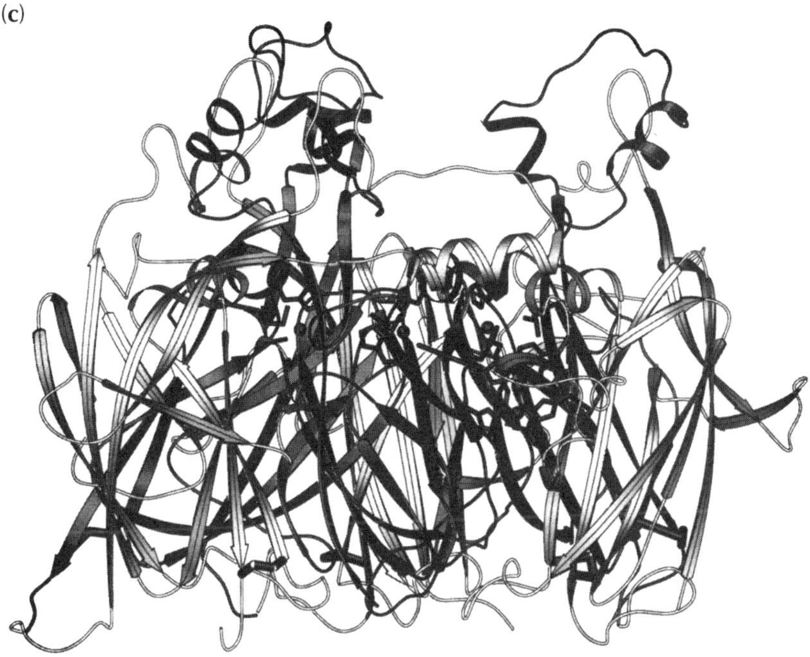

Figure 1. (*continued*) atoms are depicted as black circles and the amino acid side chains involved in copper binding are also drawn. Residues 341–346, 885–890, and 1041–1046 cannot be located unequivocally in the present electron density syntheses. There is also a chain break in domain 3 between residues 476 and 482 (on the left hand side of the figure). Interestingly this last region of ill-defined density corresponds to the cleavage site between the 67 and 50 kDa fragments. (**c**) The polypeptide chain viewed perpendicular to the pseudo threefold axis highlighting the planar nature of the bottom of the molecule and the way that the complex loops at the top of the molecule may restrict access to the mononuclear copper sites (seen in a plane near the top of the molecule). The five disulfide bridges are located near the base of the molecule.

sequences for nitrite reductase [53] ascorbate oxidase and plastocyanin [54].

The mononuclear copper atoms in domains 4 and 6 have typical type I blue copper environments with two histidine ligands, a cysteine and a methionine. The Cu–N and Cu–S(Cys) distances are around 2.1(1) Å with the Cu–S(Met) at a distance of approximately 3.0(1) Å. In the case of the domain 6 copper, the main chain carbonyl O of residue L974

(a)

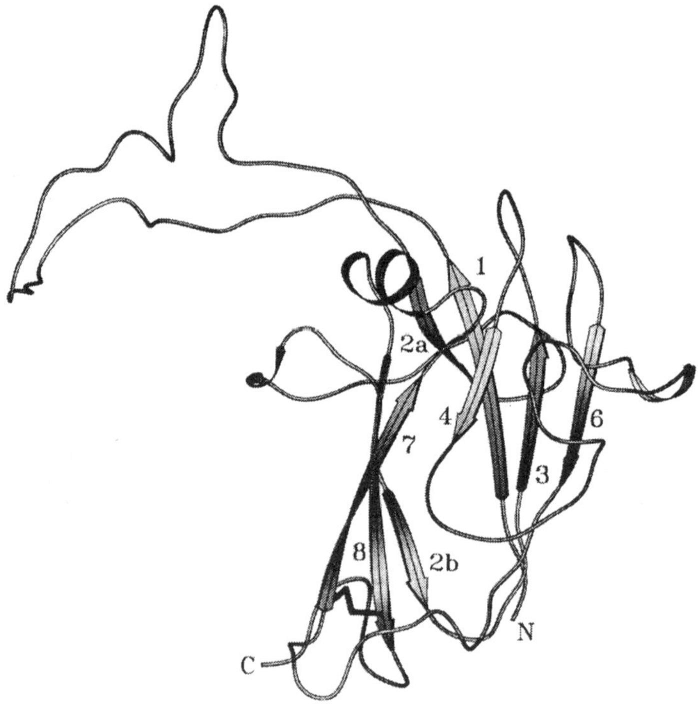

Figure 2. The cupredoxin fold in (**a**) the odd and (**b**) the even domains of hCP. The β-barrels comprise two sheets of β-strands. Sheet 1 involves strands 2a, 8, 7, and 4 and the four type I copper binding residues emanate from this sheet (see Table 3). Sheet 2 involves strands 2a, 1, 3, and 6.

appears to point towards the Cu atom as found in the oxidized form of azurin from *Alcaligenes denitrificans* [55, 56]. However, the precise orientation of this peptide group must await a higher resolution structure. In spectroscopic terms (see Table 1) the copper atoms in domains 4 and 6 are likely to be the "slow" and "fast" type I coppers, respectively, because of their distances from the trinuclear center. The mononuclear copper in domain 2 is structurally different from those in domains 4 and 6 in that it lacks the methionine residue; a nonpolar leucine residue replaces the methionine and does not appear to have any significant interaction with the copper atom. It presumably corresponds to the type IV copper in Table 1. This copper is probably in the reduced Cu(I) form in the crystal structure and indeed this may be its predominant state [57], but it is not yet clear what effect this may have on the functions of hCP.

(b)

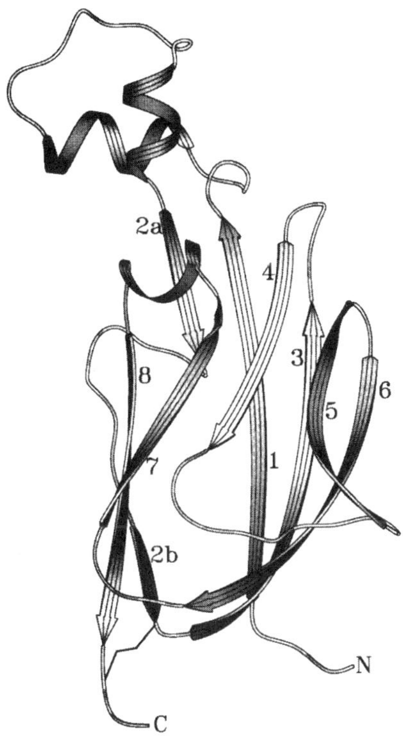

Figure 2. Continued

The disposition of the mononuclear coppers with interatomic separations of around 18 Å is intriguing; had nature organized the molecule so that the copper atoms were bound in the odd-numbered domains these intermononuclear copper distances would increase substantially since these domains point outwards from the pseudo threefold axis. Langen et al. [58] have recently studied the networks for coupling the internal redox centers in azurin and cytochrome *c* with the surface of these proteins. A distance of 18 Å between metal centers is well within the range for effective electron transfer and as shown in Figure 4 there are clear pathways linking the mononuclear copper atoms in hCP. At one end of the pathway is a hydrogen bond interaction between a Nε-atom of a histidine residue, bound to the first copper atom by its Nδ-atom, and the side chain of a conserved glutamate. The pathway then passes through an X–D–X sequence, where X is a hydrophobic residue to a second histidine attached to the second copper atom. The presence of the aspartic acid probably indicates that transfer can take place through the main

Table 2. Superposition of α-Carbon Atoms for Domains in hCP[a]

Odd Domains[b]						
	DM1	*DM3*	*DM5*	*NR1*	*AO1*	
DM3	1.0 (168)	—				
DM5	1.0 (149)	1.0 (155)	—			
NR1	1.2 (103)	1.3 (104)	1.7 (111)	—		
AO1	1.3 (110)	1.3 (110)	1.4 (106)	1.3 (102)	—	
Even Domains						
	DM2	*DM4*	*DM6*	*NR2*	*AO2*	*AO3*
DM4	0.9 (147)	—				
DM6	0.9 (143)	0.7 (144)	—			
NR2	1.7 (100)	1.4 (99)	1.5 (100)	—		
AO2	1.8 (97)	1.7 (98)	1.6 (101)	1.6 (110)	—	
AO3	1.6 (100)	1.4 (103)	1.3 (101)	1.5 (110)	1.4 (98)	—

Other Important Matches

DM2 to DM1	1.8 (91)
NR1 to DM2	1.5 (94)
NR2 to DM1	1.6 (87)

Notes: [a]All superpositions were achieved with the "lsq_improve" option in the O program suite [69].
[b][rms fit (Å) and number of atoms].

chain atoms and/or a combination of main chain and aspartate and histidine side chains. The pathways can be summarized as follows:

Pathway								
Cu6-Cu2 :	Cu6-H1026	....	E272	V273	D274	V275	H276-Cu4
Cu4-Cu6 :	Cu4-H685	....	E971	I972	D973	L974	H975-Cu6
Cu2-Cu4 :	Cu2-H324	....	E633	A634	D635	V636	H637-Cu4

In the present electron density synthesis at 3.0 Å resolution the H324(Nε) to E633 side chain distance in the Cu2 to Cu4 pathway is long

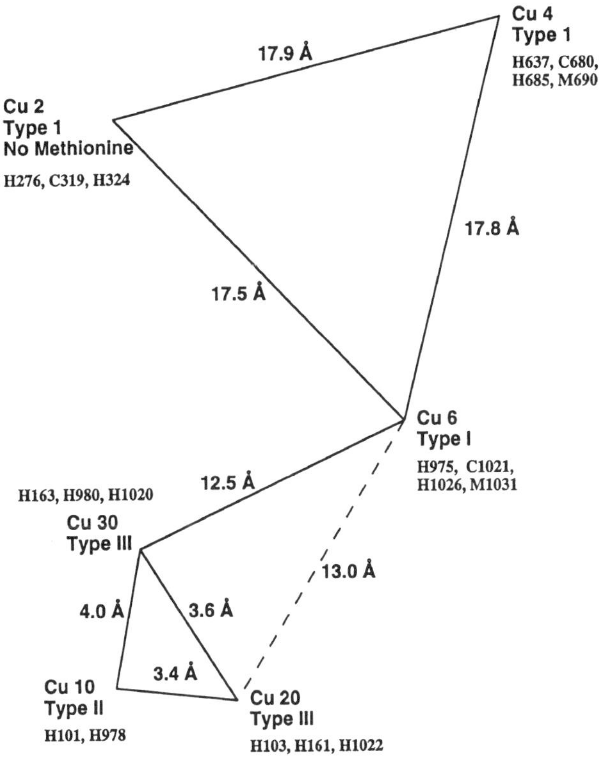

Figure 3. A schematic diagram describing the interatomic separations of the copper centers and their coordinating residues (see also Table 3.)

at 3.9 Å, compared with values of 3.0 Å for the other two pathways. Whether this is artefactual because of the limitation in resolution or a real difference implying that transfer of electrons between Cu2 and Cu4 is not significant, must await a higher resolution structure. In practice the transfer of electrons through Cu6 to the trinuclear cluster must be of prime importance, since the trinuclear center will be the site where oxygen is reduced to water. It is tempting to suggest that the difference in the nature of the domain 2 copper center from the two blue copper atoms and the differences in the electron transfer pathways are utilized by the protein depending on which of its functions it is trying to fulfill.

The trinuclear cluster is surrounded by four pairs of histidine residues with Cu–N distances of approximately 2.1(1) Å (Figure 5). Domain 1 donates two pairs H101, H103 and H161, H163 with a serine residue

Table 3. Sequence Alignment for Copper Binding Sites[a,b,c,d]

β-Strand 4: Domains which do NOT bind mononuclear copper											
					↓		↓				
DM1	98	Y	T	F	H	S	H	G	I	T	106
DM3	454	L	S	I	E	P	I	G	V	R	462
DM5	813	Y	S	I	H	A	H	G	V	Q	821
NR2	A252	T	R	P	H	L	I	G	G	H	A260
AO1	A57	V	V	I	H	W	H	G	I	L	A65
AO2	A226	A	L	N	F	A	I	G	N	H	A234
β-Strand 4: Domains which contain mononuclear copper sites											
		⇓			↓		↓				
DM2	276	H	A	A	F	F	H	G	Q	A	284
DM4	637	H	G	I	Y	F	S	G	N	T	645
DM6	975	H	T	V	H	F	H	G	H	S	983
NR1	A95	H	N	I	D	F	H	A	A	T	A103
AO3	A445	H	P	W	H	L	H	G	H	D	A453
PCY	37	H	N	I	V	F	D	.	.	.	42

β-Strands 7–8: Domains which do NOT bind mononuclear copper

										↓		↓																						
DM1	155	C	-	-	V	T	R	I	Y	H	S	H	I	-	-	D	A	P	K	D	I	A	S	-	G	L	I	G	P	L	I	I	C	181
DM3	515	C	-	-	L	A	K	M	Y	Y	S	A	V	-	-	D	P	T	K	D	I	F	T	-	G	L	I	G	P	M	K	I	C	541
DM5	855	C	-	-	I	P	W	A	Y	Y	S	T	V	-	-	D	Q	V	K	D	L	Y	S	-	G	L	I	G	P	L	I	V	C	881
NR2	A298	P	-	-	G	V	Y	A	Y	V	N	H	N	-	-	-	-	L	I	E	A	F	E	L	G	A	A	A	H	F	K	V	T	A323
AO1	A99	.	.	.	G	T	F	F	Y	H	G	H	L	-	-	G	-	-	M	Q	R	S	A	-	G	L	Y	G	S	L	I	V	D	A122
AO2	A271	N	P	S	E	N	Y	W	V	S	V	G	T	-	-	.	.	.	.	.	.	.	.	-	.	G	L	T	L	L	N	Y	L	A299

β-Strands 7–8: Domains which contain mononuclear copper sites

										↓	⇓	↓								⇓						⇓								
DM2	312	P	-	-	G	A	W	M	L	A	C	Q	N	-	-	-	-	L	N	H	L	K	A	-	G	L	Q	A	F	F	Q	V	Q	336
DM4	673	E	-	-	G	T	F	N	V	E	C	L	T	-	-	-	-	T	T	H	Y	T	G	-	G	M	K	Q	K	Y	T	V	N	697
DM6	1014	P	-	-	G	I	W	L	L	H	C	H	V	-	-	-	-	T	D	H	I	H	A	-	G	M	E	T	T	Y	T	V	L	1038
NR1	A130	.	.	.	G	V	F	V	Y	H	C	A	P	P	G	M	V	P	W	H	V	T	S	-	G	M	N	G	A	I	M	V	L	A157
AO3	A500	P	-	-	G	V	W	A	F	H	C	H	I	-	-	-	-	E	P	H	L	H	M	-	G	M	G	V	V	F	A	E	G	A524
PCY	77	P	-	-	G	T	Y	G	V	Y	C	D	P	-	-	-	-	-	-	H	S	C	A	-	G	M	K	M	T	I	T	V	.	98

Notes: [a]A dash indicates a sequence deletion, a dot indicates a structural mismatch.

[b]DM1-6, NR1-2, AO1-3 refer to domain 1, 2, 3, etc. in human ceruloplasmin, nitrite reductase [12], and ascorbate reductase [10, 11], respectively. PCY refers to the plastocyanin from green algae (Enteromorpha prolifera) [54].

[c]The sequence matches were obtained using the "lsq_improve" option in the O program suite [69].

[d]A single arrow over a column indicates residues binding to the trinuclear centers, a double arrow indicates residues binding to the type I copper sites.

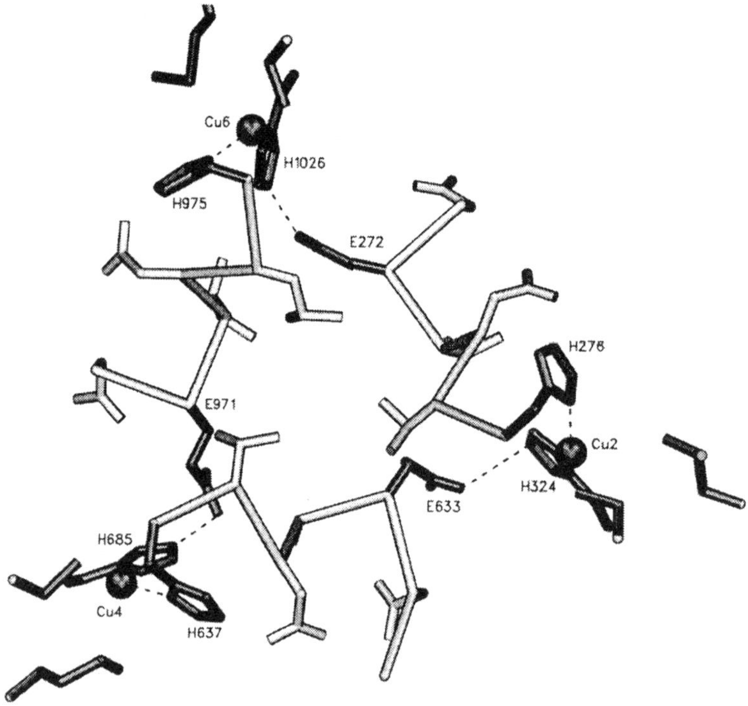

Figure 4. Electron transfer pathways between the mononuclear copper sites.

separating the components of each pair and domain 6 the remaining pairs H978, H980 and H1020, H1022, separated by phenylalanine and cysteine residues, respectively. The cysteine residue between H1020 and H1022 coordinates to the mononuclear center in domain 6. Two of the copper atoms (type III coppers) in the trinuclear center are bound by three histidines, (H103, H161, and H1022; H163, H980, and H1020), while the copper furthest away from the domain 6 copper atom is bound by only two histidine residues (H101 and H978) and is therefore designated type II; in all cases the pairs of histidine residues bridge two copper atoms. However, the electron density at the trinuclear center is consistent with the model of the trinuclear center found in AO whereby there is one oxygen atom (or a hydroxyl group) bridging the type III coppers and a second attached to the type II copper. The presence of a third oxygen atom in the middle of the trinuclear cluster can neither be discounted nor substantiated with the present resolution and quality of data. An oxygen

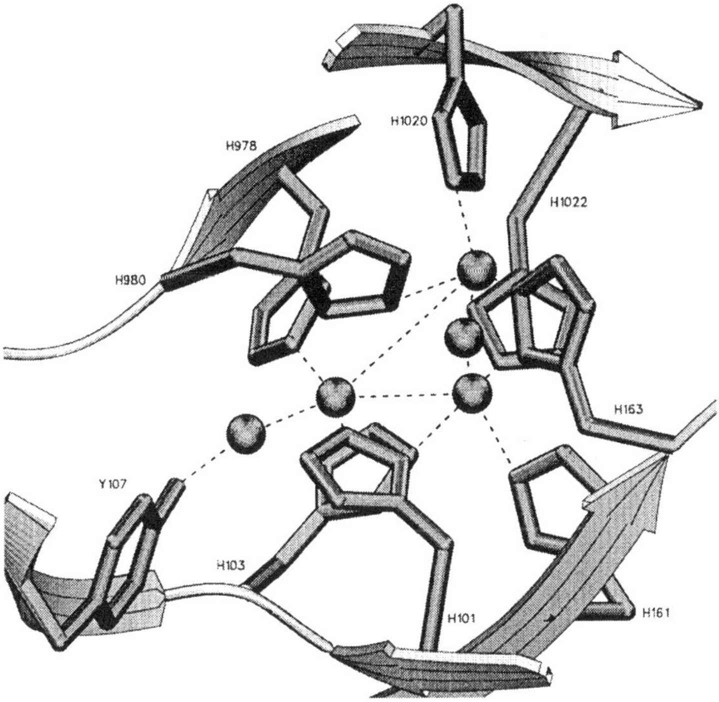

Figure 5. A diagram of the electron density at the trinuclear copper center showing the 8 coordinating histidine residues, 4 each from domains 1 and 6. An oxygen atom has been placed between the two type III copper atoms; although this fits into the observed electron density its presence is not unequivocally established. A second oxygen atom has been placed next to the type II copper; again this gives a good fit to the electron density and would participate in a hydrogen bond with the OH group of Y107.

atom attached to the type II copper gives a good fit to the electron density and is in a position to hydrogen bond to the hydroxyl group of Y107 and the main chain carbonyl oxygen of S102; in AO the OH group of the tyrosine is replaced by a water molecule which hydrogen bonds to the side chain of Q66.

In the elucidation of the X-ray structure of hCP by the method of isomorphous replacement, gold and mercury heavy atom derivatives were utilized. In the case of the mercury derivative, *p*-chloromercury-benzoate, the heavy atom bound to the free sulphydryl residue, C221, but for the gold cyanide derivative the gold atom was found to bind in the vicinity of the trinuclear copper cluster. A realistic explanation of this phenomenon is that the derivative actually binds through one of its cyanide ligands (as opposed to the gold atom itself) and in a similar way to that of azide to ascorbate oxidase. In ascorbate oxidase, two molecules of azide can bind at one of the type III copper atoms in the trinuclear center [25]. The binding of the gold derivative at the type III sites strongly suggests that, as in the case of ascorbic oxidase, the dioxygen substrate and/or oxygen-based free radicals also bind at this position for hCP.

4.3 Disulfide Bridges and Free Cysteine Residue

hCP possesses five disulfide bridges and one free cysteine residue. The free cysteine residue, C221, is located in domain 2 on the extended loop between the first and second strands and appears to be reasonably accessible as indicated by the fact that it was labeled by the mercury derivative as part of the multiple isomorphous replacement determination of the reflection phases. The disulfide bridges are located in domains 1 to 5 and serve to anchor the last strand of each domain; the C-terminal strand in domain 6 is not anchored to domain 6 through a disulfide bridge. In domains 1, 3, and 5 the disulfide bridges link strands 7 and 8 with the cysteines being separated along the chain by 25 residues; for domains 2 and 4, the disulfide bridges link strands 2b and 8 with 80 residues separating the cysteines. All five disulfide bridges are therefore located near the bottom, almost planar surface of the molecule. However, space-filling diagrams and accessibility calculations clearly indicate that, whereas the disulfide bridges in the odd domains 1, 3, and 5 are buried with very low accessibilities, those for domains 2 and 4, the copper binding domains, are highly accessible. It has been well documented that sulphydryl residues and S–S bridges make good scavengers for free radicals [59], and it is tempting to suggest that for hCP this may also be the case. That is, in addition to its capacity for reducing oxygen-based free radicals which bind to the trinuclear cluster through metal oxidation, the protein may possess alternative mechanisms for neutralizing free radicals in the serum by using its free sulphydryl residue or the S–S bridges in the copper-binding domains.

5. RELATIONSHIP WITH OTHER COPPER OXIDASES

5.1 The "Blue" Copper Oxidase Family

Ceruloplasmin is a member of the family of blue copper oxidases which also contains laccase and ascorbate oxidase. The relationship between hCP and AO [51] is quite marked and extends beyond the presence of a trinuclear copper center sited some 12–13 Å away from a mononuclear copper atom. Superposition of the three domains of AO onto the hCP structure shows that domains 1 and 3 of AO correspond to domains 1 and 6 of hCP, so that AO domain 3 and hCP domain 6 both contain a type I copper and that the trinuclear copper cluster lies at the domain interface. Furthermore, AO domain 2, although it does not bind copper, is spatially equivalent to domain 2 of hCP; that is, the subunit of AO is very similar to half the hCP pseudo hexamer. Details of these superpositions are given in Table 2 and Figure 6 shows the structural relationships schematically. It seems highly likely that the laccases which are known to contain a four-copper oxidase center have a structure essentially the same as AO[†]. Further, a superposition of the NR trimer [53] and hCP shows that the six respective domains have a similar spatial arrangement with the domains in the NR trimer, with one of the three interdomain mononuclear copper atoms in the trimer occupying a position equivalent to the trinuclear cluster in hCP. However, the intradomain coppers are bound in the three domains 1 and not domains 2. The data in Table 2 appear to indicate that domain 1 of nitrite reductase is structurally closer to domain 1 of hCP rather than to domain 2; in a similar manner domain 2 of NR appears structurally more similar to domain 2 of hCP than domain 1, despite the reversal in copper-binding capability.

Evolutionary pathways for the cupredoxins and multi-copper oxidases have been described by Rydén and Hunt [60]. It seems likely that AO, hCP, and NR have all evolved from a two domain ancestral protein (gene duplication of a cupredoxin) which had type I coppers in both domains. As indicated in Figure 6, one possibility is that the precursor consisted of domains equivalent to 1 and 2 in the hCP structure; the conservation of hydrogen-bonding and ion-pair patterns between domains 1 and 2, 3, and 4, and 5 and 6, respectively, in hCP appears to support this hypothesis. From such a precursor, loss of copper-binding capability in domain 1, followed by the insertion of a spacer domain would enable the formation of a three-domain subunit as found in AO. In this three-domain unit the original domain 2 would be equivalent to domain 3 in Figure 6 and the formation of a trinuclear binding site at the interface of domains 1 and 3 would be the last stage in the evolutionary pathway. In the case of hCP gene triplication of the precursor (with copper only in domain 2) would lead to a six-domain protein with coppers in the even domains. Again the formation of a trinuclear-binding site between domains 1 and 6 (equivalent to AO domains 1 and 3) would be a final stage in the

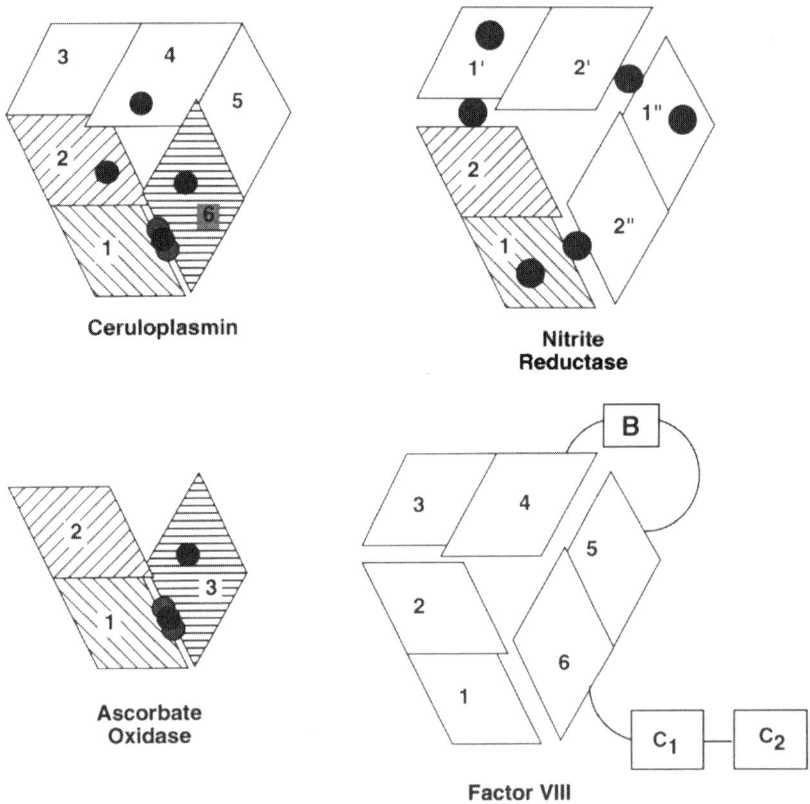

Figure 6. Structural relationships between ascorbate oxidase, ceruloplasmin, nitrite reductase, and blood clotting factor VIII.

evolution of the protein. On the other hand, the loss of copper-binding capability from domain 2 of the two-domain precursor, followed by trimerization would be one stage leading to the NR structure. Insertion of the interdomain mononuclear copper sites would be a final stage. An obvious alternative to the above scheme involves a two domain precursor with an additional copper-binding site at the domain interface.

It is interesting to speculate why nitrite reductase has its type I coppers in domains 1, whereas in hCP the mononuclear copper binding sites are retained in the domains 2, 4, and 6 where they are comparatively buried in the protein. One possible reason can be related to the difference in functions of the two proteins. NR has to interact with a relatively large pseudo-azurin macromolecule in order for electron transfer to take place, whereas on the other hand, hCP probably functions through interactions with small molecules. hCP may achieve this aim by placing the coppers towards the inside of the macromolecule and having substantial loop regions which restrict access to the mononuclear copper sites.

5.2 Blood Clotting Factor VIII

Sequence homology has suggested that ceruloplasmin and the blood clotting factors V and VIII constitute a family of structurally related proteins [6] and as shown in Figure 6 the hCP structure can be readily used to formulate a putative model for the blood clotting factor VIII [61]. The A_1, A_2, and A_3 domains of factor VIII can be considered equivalent to hCP domains 1 and 2, 3 and 4, and 5 and 6, respectively; i.e. each A domain consists of two hCP domains. The C_1 and C_2 domains of factor VIII form a C-terminal tail which may act as a membrane anchor. The large B domain of factor VIII is excised during activation of the coagulation factor and appears to be connected to the two A domains at their N-terminal ends; thus it could form a protective cover over the relatively planar, bottom surface of the molecule. In the model this surface would contain two peptides which are known to participate in the binding of factor VIII to factor IXa. This model [61] and further details of the structural relationships in the blue copper oxidase family of protein will be pursued elsewhere.

Prior to the determination of the structure of hCP, a homology modeling study of the triplicated A domains of factor VIII had been undertaken using the structure of nitrite reductase [62]. This model was built on the premise that domain 1 of NR was aligned with the even domains of factor VIII, partly because the sequence alignment between NR domain 1 and the factor VIII even domains appeared slightly better than that for the odd domains, and partly because factor VIII contains two conserved sets of potential mononuclear copper binding sites, one in domain 2 and one in domain 6, and it was thought that these would be equivalent to the copper-binding domains in NR. As indicated by Pan et al. [61], aligning domain 1 of NR with the factor VIII odd domains would yield a structure similar only in general appearance. Figure 6 clearly indicates, however, that this latter alignment should be the case; i.e. the inverse of the model based on NR, and accordingly model proposed by Pan et al. should be viewed with caution since it is liable to be incorrect in all but the broadest features.

6. CERULOPLASMIN, IRON METABOLISM, AND COPPER TRANSPORT

6.1 Ceruloplasmin and Systemic Hemosiderosis

As indicated in Section 1 the view that ceruloplasmin is intricately linked with iron metabolism has recently received renewed attention. Recent papers [39,40,63] associating a hereditary ceruloplasmin deficiency with aceruloplasminemia and severe iron deposition in visceral organ and brain tissues have focused attention on the putative role of ceruloplasmin as a ferroxidase. In certain cases, systemic hemosiderosis appears to be caused by incomplete expression of the hCP molecule, with the polypeptide chain being prematurely terminated. These observations gives credence to the proposition that the enzyme may be involved in facilitating the release of intracellular iron and its oxidation prior to uptake and transport by transferrin. Further, the X-ray structure provides structural explanations as to why, if the truncated molecules are actually secreted into the bloodstream, they will have their oxidase function severely impaired and may be highly susceptible to proteolytic cleavage. Thus, Figure 7 shows the trace of the α-carbon polypeptide backbone for domains 1 and 6 of human ceruloplasmin only; domain 1 is shaded grey on the left hand side of the figure. The full and unshaded structures on the right hand side of the figure represent domain 6 and, in particular, the full trace is the portion of the polypeptide chain, residues 991 to 1046 inclusive, which would be missing if incorrect expression terminated the chain at 991 [39,63]. The copper-binding ligands that would be concomitantly missing comprise three of the ligands that bind to the domain 6 mononuclear copper (C1021, H1026, and M1031). In the absence of these ligands, it can be concluded that domain 6 is very unlikely to bind copper and this may have a profound effect on the oxidase efficiency of the enzyme. In addition, two of the histidine residues that bind to the trinuclear cluster (H1020 and H1022) are also absent which gives rise to the possibilities that this cluster may not bind in the same way as in the intact enzyme or that it may not bind at all; in either case this will affect the oxidase activity of the enzyme. Further, the amino acid residues on the missing peptide contribute over 50% of the hydrogen-bond and ion-pair interactions between domains 1 and 6 in the intact enzyme. It is therefore highly probable that the incomplete enzyme will adopt a different organization of the tertiary structure and may well exist in an

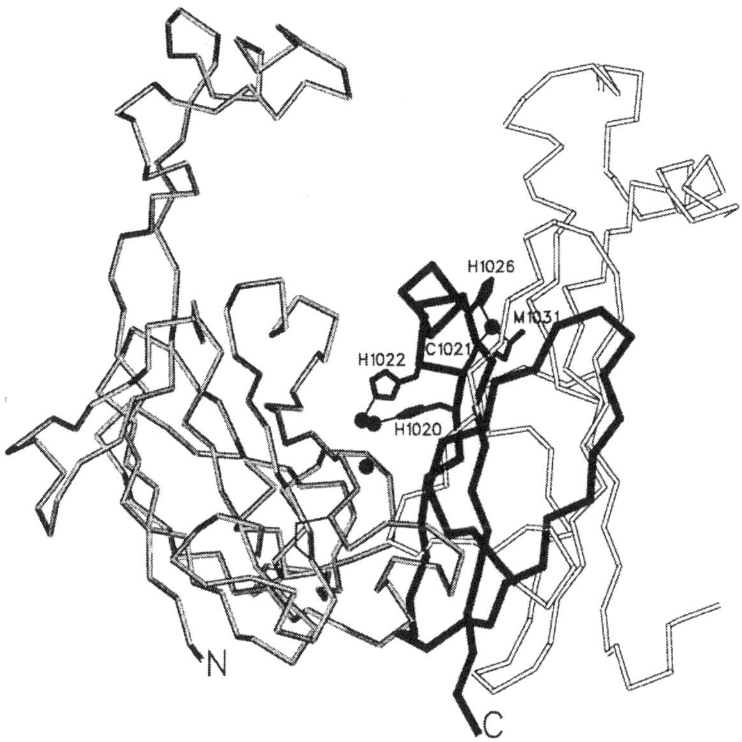

Figure 7. Traces of the α-carbon polypeptide backbone of domains 1 and 6 in the hCP structure. Domain 1 is shown (shaded) on the left hand side of the diagram; this domain contributes four histidine residues (not shown) to the trinuclear cluster; copper atoms are depicted as black spheres. Domain 6 is on the right hand side of the figure and also contributes four histidine residues to the cluster. The portion of the polypeptide chain colored black is that which is missing in the truncated enzyme. This polypeptide, residues 991 to 1046 inclusive, includes two histidine residues bound to the trinuclear copper center and three residues bound to the mononuclear copper in domain 6; these residues are depicted in black. The absence of the C-terminal polypeptide would also remove over 50% of the interdomain hydrogen-bond and iron-pair interactions observed in the intact enzyme.

open configuration. Such a configuration will be even more susceptible to proteolytic cleavage than the correctly folded enzyme.

The net effect of the missing C-terminal portion of the polypeptide chain is that the enzyme may not fold into the organizational tertiary structure depicted schematically in Figure 1 and that its ability to act as an oxidase will be severely impaired with respect to the intact enzyme. Incomplete folding and increased susceptibility to proteolytic cleavage would be consistent with the observed aceruloplasminemia such as the failure to find ceruloplasmin in liver extracts by the Western blot and ELISA methods. An impairment of the ferroxidase activity would be consistent with a gradual accumulation of intracellular iron leading to systemic hemosiderosis; the truncated enzyme would not mediate the release of iron from cells nor its oxidation prior to uptake by transferrin. Truncation of the ceruloplasmin chain at residue 858 [40], that is, with whole of domain 6 and a portion of domain 5 missing, would be even more catastrophic. The secretion of this protein into the bloodstream would almost certainly lead to swift degradation and complete loss of ferroxidase activity.

6.2 Putative Ferroxidase Sites

If the protein has genuine ferroxidase activity it should be possible to identify putative metal binding sites where electrons can be readily donated to the domain 6 mononuclear copper, either directly, or through one of the electron transfer pathways indicated in Section 4. In the case of AO, the organic substrate is believed to bind in the vicinity of the mononuclear copper adjacent to the trinuclear center. It has been proposed that one of the histidine residues bound to the copper by its Nδ atom has the Nε atom available for substrate interaction and that the site is bounded by two tryptophan residues [51]. Neither of these tryptophan residues is conserved in hCP, but in domains 4 and 6 of hCP one of them is replaced by a histidine as shown in Figure 8. In domain 6 this histidine, H940, is only some 3–4 Å away from H975 at the copper 6 binding site. Furthermore H940 is in close proximity to three negatively charged residues, E935 and D1025 from domain 6 and E272 from domain 2. Indeed, in the current model a strong electron density peak is observed which can be interpreted as a metal atom coordinated to these four residues, and possibly to a water molecule. In domain 4 the situation is similar (H602, E597, D684, and E971 from domain 6), but less well-defined indicating some disorder at the putative metal binding site.

The nature of the metal cannot be determined unambiguously from the electron density alone and rigorous metal assays of hCP crystals using X-ray fluorescence and atomic absorption techniques are in progress. However, from the blue color of the crystals and the nature of the ligands,

(**a**)

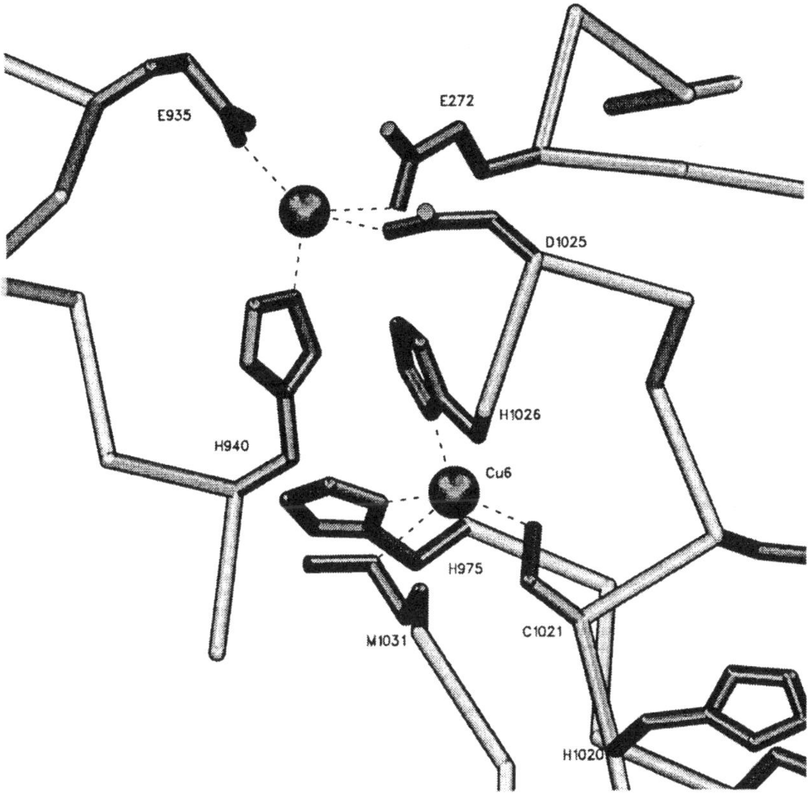

Figure 8. The hCP molecule possesses putative cation binding centers in the vicinity of the mononuclear copper in atoms in domains 4 and 6. These centers could be used to bind Cu(II) for specific copper transport, or Fe(II) for ferroxidase activity. The centers comprise a histidine residue and up to three acidic residues are relatively accessible. At 3.0 Å resolution it is not possible to define these sites in detail and water molecules may also be involved. (**a**) Electron transfer can readily take place from the cation to the domain 6 copper via histidine residues H940 and H1026 or alternatively through D1025 and H1026. (**b**) The cation binding site in this domain appears less ordered than domain 6, but could still involve H602 and up to three acidic residues, E597, D684, and E971. Electron transfer from the cation to the domain 4 copper would be followed by transfer to the domain 6 copper (see Figure 4). (**c**) In the equivalent region in domain 2 the histidine residue is replace by a tyrosine. The electron density shows no evidence of cation binding. (**d**) The relationship of the cation binding site in domain 6 to the trinuclear copper center.

(b)

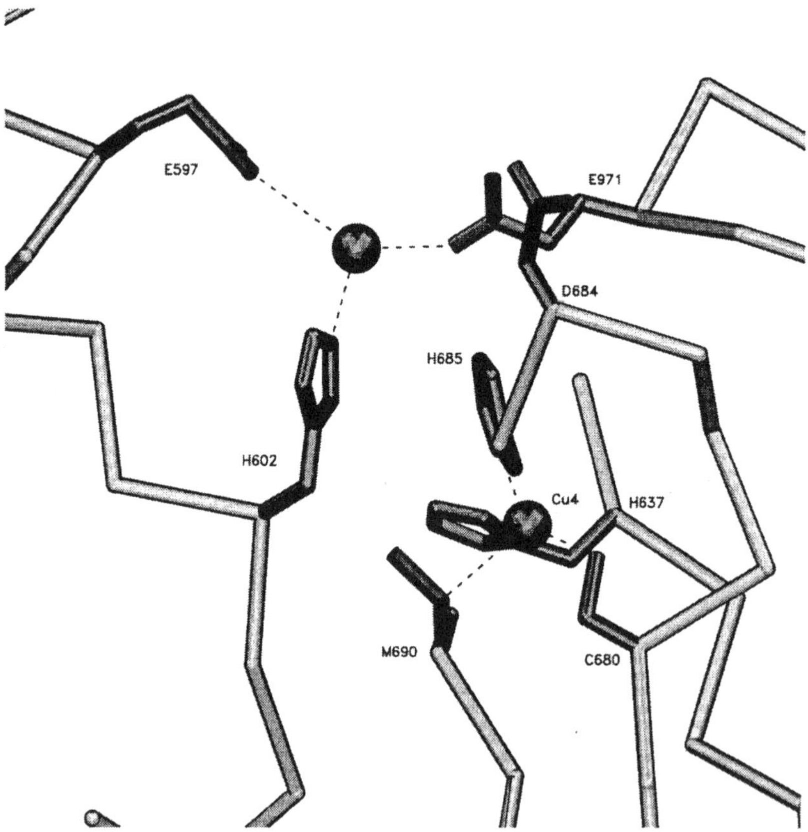

Figure 8. Continued

extra copper is a distinct possibility; the heights of the electron density maxima are approximately 50% of the main copper peaks, indicating incomplete occupancy. It is also possible that these sites can bind other transition metals and very recent soaking experiments on the crystals with $CoCl_2$ have shown the presence of cobalt. Binding of ferrous iron would readily facilitate electron donation to the trinuclear center with concomitant oxidation of the metal to ferric iron with an oxidation state matching the three negatively charged residues. In the case of domain 6 one possibility for this transfer would be through D1025 and H1026 (these two residues are within hydrogen-bonding distance) to the mononuclear copper atom, Cu6, and then to the trinuclear center through

(c)

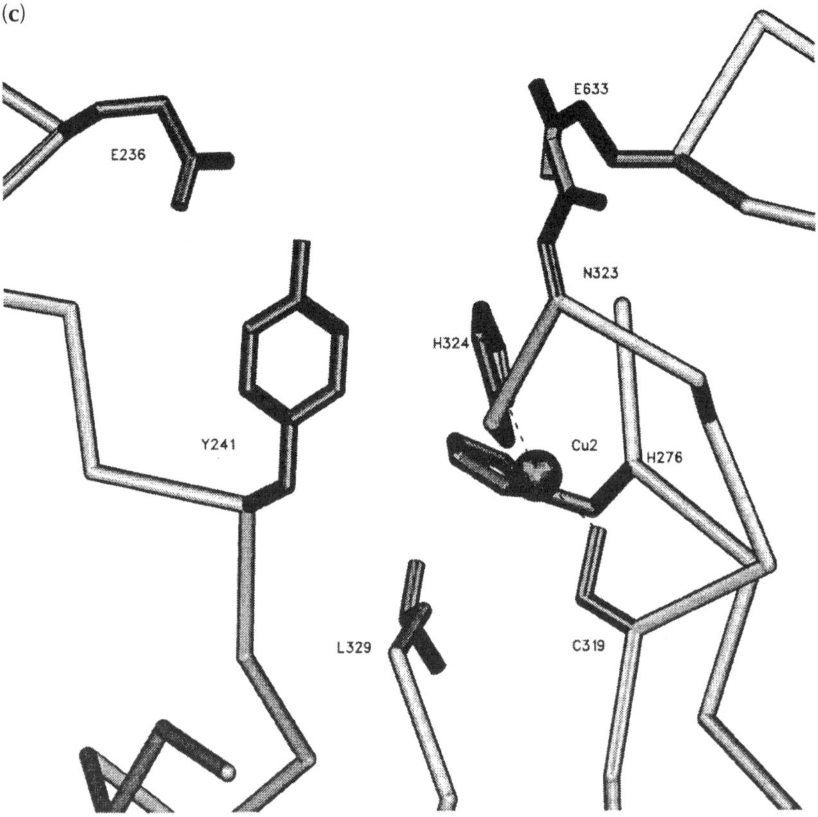

Figure 8. Continued

C1021 and H1020, and/or H1022. The situation in domain 2 is once again different. In this domain a tyrosine residue, Y241, replaces the histidine, and other pertinent residues are E236, N333, and E633 from domain 4; there appears no significant peak in the electron density synthesis. Whether metals and/or organic substrates bind to this site requires further research.

It therefore appears that for at least domains 4 and 6, hCP possesses metal binding sites in addition to the six integral copper atoms. Binding of ferrous iron at these sites could result in its oxidation and its removal from the plasma or in the reduction of oxygen-based free radicals. Binding of copper could result in its transport through the plasma (*vide infra*).

(d)

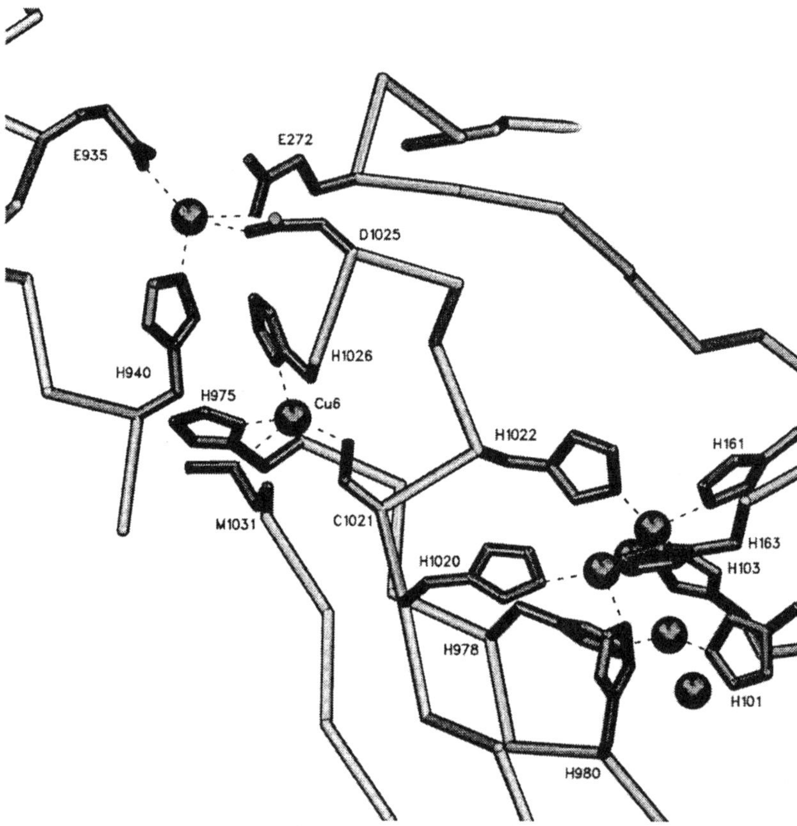

Figure 8. Continued

6.3 Copper Transport

Delivery of copper may take place by a natural turnover mechanism for ceruloplasmin in the plasma through which copper ions move from the major sites of ceruloplasmin synthesis in liver cells to peripheral tissues for incorporation into copper dependent enzymes. However, this mode of delivery alone contrasts strongly with that of the serum iron transporting protein, transferrin. The latter is a molecule comprising two structurally homologous lobes with each lobe containing two dissimilar domains and an iron-binding site located deep within the interdomain cleft [64]. In the *apo*-protein, the lobes have a conformation in which the domains can be wide open, but on binding ferric iron, the domains close to protect the bound metal. The iron-loaded protein is recognized by cell surface receptors, principally in the bone marrow, and the iron delivered to cells by receptor-mediated endocytosis, followed by release of the transferrin back into the bloodstream. It is difficult to envisage similar mechanisms for copper delivery by ceruloplasmin, but as indicated in the previous section, there is some evidence that the protein can bind up to two additional copper atoms in relatively accessible sites. It is these copper atoms that could be utilized in copper transport in a manner independent of protein turnover. Further studies are clearly necessary.

7. FUTURE

The X-ray analysis of human ceruloplasmin at 3.0 Å resolution has provided an important template which can be used to interpret biochemical and spectroscopic studies and to begin to elucidate the "enigma" regarding the precise functions of ceruloplasmin. Clearly, there are many questions still to be answered, but extensions of the X-ray crystallographic approach have the potential to provide valuable new information in a number of ways as outlined below. However, a crucial issue that will inevitably permeate all future experiments will be the resolution of the data. A structure at a resolution appreciably higher than 3.0 Å, and preferably better than 2.0 Å, will be essential if details of the molecular structure are to be precisely defined. Such details include the disposition of the oxygen (hydroxyl) atoms in the trinuclear cluster and the mode of binding of anions to this metal center. A number of possibilities are available to achieve high-resolution data, including:

1. *Cryogenic techniques*. Flash-freezing protein crystals to near liquid nitrogen temperature has become a routine method for minimizing radiation damage during X-ray data collection. It also frequently leads to higher resolution data since the thermal motions of individual atoms are greatly reduced and highly flexible regions of the polypeptide chain at ambient temperatures are locked into a particular conformation (or discrete conformations) at low temperature. The technique involves partially replacing the solvent by cryoprotectants [65], so that on freezing the water forms a vitreous amorphous phase which does not severely disrupt the crystal lattice.
2. *Cocrystallization*. Cocrystallization of hCP with anion and/or cation inhibitors may stabilize the oxidation states of the copper atoms, prevent loss of copper, and therefore stabilize the overall

conformation of the protein, leading to higher resolution diffraction data.

3. *Modification of glycan chains.* Preparations of hCP exhibit heterogeneity, particularly with respect to the glycan moieties and at least two heteroforms are present in solution. Only the first one or two sugar residues of three of the four possible glycan chains can be observed in the current electron density syntheses at 3.0 Å resolution, indicating that these chains are not localized and are probably disordered in the solvent regions of the crystals. Whether they have any influence on the crystal packing is questionable, but separation of the heteroforms using, for example, column chromatography on hydroxyapatite, will lead to hCP preparations with a higher "crystallographic" grade of purity and crystallization of the separated isoforms could lead to better diffracting crystals. Partial or complete enzymatic removal of the glycan chains may also be possible and this could also lead to better quality crystals and therefore higher resolution diffraction data. Such experiments may also help to define roles for the glycan chains in hCP.
4. *Ceruloplasmin from other sources.* Alternative approaches could involve the isolation of ceruloplasmin from other sources yielding crystals with a tighter molecular packing and higher resolution diffraction patterns.

Although the main thrust of future crystallographic work must be oriented towards a higher resolution structure, many further experiments can be envisaged which should help to probe structure–function relationships and physiological roles for hCP in the plasma. Such experiments include:

1. *Loss of copper.* Solutions and crystals of ceruloplasmin that have been allowed to stand for several weeks partially lose their blue color and this may indicate changes in copper oxidation state and/or loss of copper. A series of experiments are required to determine whether any copper loss is site preferential and how this is affected by anionic inhibitors.
2. *Binding of anions, cations, and organic substrates.* In order to probe the roles of the different types of copper atoms and the nature of the electron transfer pathways between them, a variety of complexes of hCP with inhibitors of oxidase activity and

pseudo-substrates need to be studied. These complexes include anions (azide, fluoride, cyanide, and chloride), divalent cations (Fe, Cu, Co, Ni, and Zn), and synthetic analogues of *p*-phenylenediamine and other biogenic amines and phenols. As indicated in Section 6, soaking hCP crystals with $CoCl_2$ has already provided supporting evidence for ferroxidase activity and/or a possible copper transport function. These experiments can be undertaken either by soaking the hCP crystals in solutions containing the substrate or by cocrystallization.

3. *Site-directed mutagenesis.* At the present time it is unclear whether ceruloplasmin can be expressed and correctly folded using recombinant methods, although preliminary investigations are in progress using both and mammalian yeast cell lines [66]. Provided that these problems can be surmounted, the molecular structure clearly suggests a number of key site-directed mutagenesis studies, particularly those involving residues bound to the copper atoms. For example, it would be interesting to mutate away the copper binding sites in domains 2 and 4 and examine the efficiency of catalysis and range of substrates handled by the modified ascorbate oxidase-like protein. The domain 2 copper environment could be modified by changing the leucine to a methionine and/or the domain 4 and 6 coppers could have the methionine changed to a leucine. The permutations and combinations seem unlimited and the most direct probes of the functions and mechanisms of ceruloplasmin may only become apparent after the more obvious possibilities have been explored.

Ceruloplasmin, in common with most of the plasma proteins and components, provides a fascinating field of study and merits extensive further investigation, particularly with regard to its involvement in iron metabolism and its relationship with the blood clotting factors V and VIII. It is highly likely that the protein is genuinely multifunctional and indeed may serve as a "molecular vacuum cleaner" in the blood, assisting in the mobilization of iron and ensuring that no free ferrous iron can give rise to free radical damage, counteracting the possibilities of other oxygen free radical damage, oxidizing biogenic amines, and generally removing potentially harmful small-molecule chemical reagents from the environment. This review has been entitled "Ceruloplasmin: The Beginning of the End of an Enigma"; undoubtedly further studies will progress beyond this point.

ACKNOWLEDGMENTS

The authors gratefully acknowledge the support of the Medical Research Council (for VNZ), the Biotechnology and Biological Sciences Research Council (for a studentship to GC) and the Engineering and Physical Sciences Research Council. The structure analysis of ceruloplasmin was undertaken as part of a collaborative research program at Daresbury Laboratory supported by all three Research Councils under the auspices of the Joint Biology Programme Steering Committee. The support of the Council for the Central Laboratories of the Research Councils is also acknowledged and in particular the Protein Crystallography Project Team and the Biology Support Staff at the SRS, Daresbury Laboratory. We would also like to thank the Royal Society (UK) for providing travel opportunities between the UK and Russia (for VNZ and PFL), the late Professor B.Vainshstein (Institute of Crystallography, Moscow) for his support and encouragement and Professor Elinor Adman for many interesting discussions and suggestions. During the structure analysis, considerable use was made of the CCP4 package of crystallographic programs [67] and many of the figures were drawn using the program SETOR [68].

NOTE

†This has now been shown to be the case. Ducros, V., Brzozowski, A.M., Wilson, K.S., Brown, S.H., Østergaard, P., Schneider, P., Yaver, D.S., Pedersen, A.H. and Davies, G.J. Nature Struct. Biol., 5 (1998) 310–316.

REFERENCES

[1] Linder, M.C., in *Biochemistry of Copper*, Ch. 1, Plenum Press, New York, 1991, pp. 1–13.

[2] Holmberg, G.C., Acta Physiol. Scand., 8 (1944) 227–229.

[3] Laurie, S.N. and Mohammed, E.S., Coord. Chem. Rev., 33 (1980) 273–313.

[4] Frieden, E. and Hsieh, H.S., in Meister, A. (ed.), *Advances in Enzymology and Related Areas in Molecular Biology*, Vol. 44, John Wiley & Sons, New York, 1976, pp. 187–236.

[5] Saenko, E.L., Yaropolov, A.I. and Harris, E.D., J. Trace Elements in Exp. Med., 7 (1994) 69–88.

[6] Church, W.R., Jerningan, R.L., Toole, J., Hewick, R.M., Knopf, J., Knutson, G.J., Nesheim, M.E., Mann, K.G. and Fass, D.N., Proc. Natl. Acad. Sci. USA, 81, (1984) 6934–6937.

[7] Zaitseva, I., Zaitsev, V., Card, G., Moshkov, K., Bax, B., Ralph, A. and Lindley, P.F., J. Biol. Inorg. Chem., 1 (1996) 15–23.

[8] Dwulet, F.E. and Putnam, F.W., Proc. Natl. Acad. Sci. USA, 78 (1981) 2805–2809.

[9] Takahashi, N., Bauman, R.A., Ortel, T.L., Dwulet, F.E., Wang, C.-C. and Putnam, F.W. Proc. Natl. Acad. Sci. USA, 80 (1983) 115–119.

[10] Messerschmidt, A., Rossi, A., Ladenstein, R., Huber, R., Bolognesi, M., Gatti, G., Marchenisi, A., Pertuzelli, R. and Finazzi-Agró, A., J. Mol. Biol., 206 (1989) 513–529.
[11] Messerschmidt, A. and Huber, R. Eur. J. Biochem., 187 (1990) 341–352.
[12] Fenderson, F.F., Kumar, S., Adman, E.T., Lui, M.-Y., Payne, W.J. and LeGall, J., Biochem., 30 (1991) 7180–7185.
[13] Linder, M.C., in *Biochemistry of Copper*, Ch. 4, Plenum Press, New York, 1991, pp. 73–134.
[14] Reinhammer, B. and Malmström, B.G., in Spiro, T.G. (ed.), *Copper Proteins*, John Wiley & Sons, New York, 1981, pp. 111–149.
[15] Frieden, E., in Sigel, H. (ed.), *Metal Ions in Biological Systems*, Marcel Dekker, New York, 1981, pp. 117–142.
[16] Van Leeuwen, F.X.R. and Van Gelder, B.F., Eur. J. Biochem., 87 (1978) 305–312.
[17] Evans, R.W., Madden, A,D., Patel, K.J., Gibson, J.F. and Wrigley, K.S., Biochem. Soc. Trans., 13 (1985) 627–629.
[18] Rydén, L., in Lontie, R. (ed.), *Copper Proteins and Copper Enzymes*, Vol. 3, CRC Press, 1984, pp. 37–101.
[19] Solomon, E.I., Lowery, M.D., LaCroix, L.B. and Root, D.E., in Riordan, J.F. and Vallee, B.L. (eds.), *Methods in Enzymology*, Vol. 226, Academic Press, 1993, Ch. 1, pp. 1–33.
[20] Curzon, G. and Cummings, J.N., in Peisach, J., Aisen, P. and Blumberg, W.E. (eds.), *The Biochemistry of Copper*, Academic Press, New York, 1966, pp. 545–557.
[21] Andreasson, L.E. and Vänngård, T. Biochem. Biophys. Acta, 200 (1970) 247–257.
[22] Dawson, J.H., Dooley, D.M. and Gray, H.B., Proc. Natl. Acad. Sci. USA, 75 (1983) 4078–4081.
[23] Byers, W., Curzon, G., Garbett, K., Speyer, B.E., Young, S.N. and Williams, R.J.P., Biochim. Biophys. Acta, 310 (1973) 38–50.
[24] Manabe, T., Manabe, N., Hiromi, K. & Hatano, H., FEBS Lett., 16 (1971) 201–203.
[25] Messerschmidt, A., Luecke, H. and Huber, R., J. Mol. Biol., 230 (1993) 997–1014.
[26] Ando, K. J., Biochemistry (Tokyo), 68 (1970) 501–508.
[27] Musci, J., Bonaccorsi di Patti, M.C. and Calabrese, L.J., J. Prot. Chem., 14 (1995) 611–619.
[28] Wilson, S.A.K., Brain, 34 (1912) 295–509.
[29] Menkes, J.H., Alter, M., Steigleder, G., Weakley, D.R. and Sung, J.H., Pediatrics, 29 (1962) 764–779.
[30] Danks, D.M., Campbell, P.E., Walker-Smith, J., Stevens, B.J., Gillespie, J.M., Blomfield, J. and Turner, B., The Lancet, i (1972) 1100–1102.
[31] Danks, D.M., Campbell, P.E., Stevens, B.J., Maynes, V. and Cartwright, E., Pedriatics, 50 (1972) 188–201.
[32] Klevay, L.M., in Draper, H.H. (ed.), *Advances in Nutritional Research*, Vol. 1, Plenum Press, New York, 1977, pp. 227–252.
[33] Linder, M.C., in *Biochemistry of Copper*, Ch. 9, Plenum Press, New York, 1991, pp. 331–366.

[34] Frydman, M., Bonne-Tamir, B., Farrer, L.A., Conneally, P.M., Magazanik, A., Ashel, A. and Goldwitch, Z., Proc. Natl. Acad. Sci. USA, 82,
[35] Young, S.N. and Curzon, G., Biochem. J., 129 (1972) 273–283.
[36] Gunarsson, P.O., Nylen, U. and Pettersson, G., Eur. J. Biochem., 37 (1973) 41–46.
[37] Eggert, C., Temp, U., Dean, J.F.D. and Eriksson, K.-E.L., FEBS Letts., 376 (1995) 202–206.
[38] Osaki, S., Johnson, D.A. and Frieden, E., J. Biol. Chem., 241 (1966) 2746–2751.
[39] Yoshida, K., Furihata, K., Takeda, S., Nakamura, A., Yamamoto, K., Morita, H., Hiyamuta, S., Ikeda, S., Shimizu, N. and Yanagisawa, N., Nature Genetics, 9 (1995) 267–272.
[40] Takahashi, Y., Miyajima, H., Shirabe, S., Nagataki, S., Suenaga, A. and Gitlin, J.D., Human Mol. Genet., 5 (1996) 81–84.
[41] Gutteridge, J.M.C., Chem.-Biol. Interact., 56 (1985) 113–120.
[42] Cousins, R.J., Physiol. Rev., 65 (1985) 238–309.
[43] Harris, E.D., Proc. Soc. Exp. Biol. Med., 196 (1991) 130–140.
[44] Owen, C.A., Jr., Amer. J. Physiol., 209 (1965) 900–904.
[45] Owen, C.A., Jr. and Hazelrig, J.B., Amer. J. Physiol., 210 (1966) 1059–1064.
[46] Neifakh, S.A., Monakhov, N.K., Shaposhnikov, A.M. and Zubzhitski, Y.N., Experientia, 25 (1969) 337–344.
[47] Shaposhnikov, A.M., Zubzhitski, Y.N. and Shulman, V.S., Experientia, 25 (1969) 424–426.
[48] Hilton, M., Spenser, D.C., Ross, P., Ramsey, A. and McArdle, H.J., Biochim. Biophys. Acta, 1245 (1995) 153–160.
[49] Adman, E.T., Advances in Prot. Chem., 42 (1991) 145–197.
[50] Ortel, T.L., Takahashi, N., Bauman, R.A. and Putnam, F.W., Protides Biol. Fluids, 31 (1983) 243–248.
[51] Messerschmidt, A., Ladenstein, R., Huber, R., Bolognesi, M., Avigliano, L., Petruzzelli, R., Rossi, A. and Finazzi-Agró, A., J. Mol. Biol., 224 (1992) 179–205.
[52] Malmström, B.G., Ann. Rev. Biochem., 51 (1982) 21–59.
[53] Godden, J.W., Turley, S., Teller, D.C., Adman, E.T., Liu, M.-Y., Payne, W.J. and LeGall, J., Science, 253 (1991) 438–442.
[54] Collyer, C.A., Guss, J.M., Sugimura, Y., Yoshizaki. F. and Freeman, H.C., J. Mol. Biol., 211 (1990) 617–632.
[55] Baker, E.N., J. Mol. Biol., 203 (1988) 1071–1095.
[56] Nar, H., Messerschmidt, A., Huber, R., van der Kamp, M. and Canters, G.W., J. Mol. Biol., 221 (1991) 765–772.
[57] Calabrese, L. and Leuzzi, U., Biochem. Intl., 8 (1984) 35–39.
[58] Langen, R., Chang, I.-J., Germanas, J.P., Richards, J.H., Winkler, J.R. and Gray, H.B., Science, 268 (1995) 1733–1735.
[59] Thomas, J.A., Poland, B. and Honzatko, R., Arch. Biochem. Biophys., 319 (1995) 1–9.
[60] Rydén, L.G. and Hunt, L.T., J. Mol. Evol., 36 (1993) 41–66.
[61] Pemberton, S., Lindley, P., Zaitsev, V., Card, G., Tuddenham, E.G.D. and Kemball-Cook, G., Blood, 89 (1997) 2413–2421.

[62] Pan, Y., DeFay, T., Gitschier, J. and Cohen. F.E., Nature Struct. Biol., 2 (1995) 740–744.
[63] Morita, H., Ikeda, S., Yamamoto, K., Morita, S., Yoshida, K., Nomoto, S., Kato, M. and Yanagisawa, N., Ann. Neurology, 37 (1995) 646–656.
[64] Baker, E.N. and Lindley, P.F., J. Inorg. Biochem., 47 (1992) 147–160.
[65] Mitchell, E.P. and Garman, E.F., J. Appl. Cryst., 27 (1994) 1070–1074.
[66] Evans, R.W. Private communication.
[67] CCP4 Suite: Programs for Protein Crystallography. Acta Cryst., D50 (1994) 760–763.
[68] Evans, S.V., J. Mol. Graphics, 11 (1993) 134–138.
[69] Jones, T.A., Zou, J.-Y., Cowan, S.W. and Kjeldgaard, M., Acta Cryst., A47 (1991) 110–119.

THE CHEMISTRY OF RHENIUM IN NUCLEAR MEDICINE

Philip J. Blower and Sushumna Prakash

Perspectives on Bioinorganic Chemistry
Volume 4, pages 91–143.

ISBN: 0-7623-0352-2

1. INTRODUCTION

The modern clinical discipline of nuclear medicine comprises two main divisions: the application of γ- and positron-emitting radionuclides for the imaging, and hence diagnosis, of disease; and the application of α- and β-emitting (and Auger-electron emitting) radionuclides for cancer-selective cytotoxic therapy, treatment of arthritis, and the prevention of restenosis after coronary angioplasty. These applications exploit physiological and biochemical pathways to achieve selective delivery of radionuclides to target tissues following systemic (usually intravenous) administration.

The radioisotopes most heavily used in nuclear medicine are selected primarily on the basis of their nuclear properties and the economics of their manufacture and distribution, rather than their intrinsic biological relevance. Isotopes with near-ideal properties for imaging are relatively few. The principal diagnostic radionuclide in modern nuclear medicine is an isotope of technetium (^{99m}Tc), an element that does not exist naturally, having no stable isotopes. The range of potentially useful therapeutic radionuclides is wider, and the prevalence of metallic elements in the list provides inorganic chemists with a significant challenge: to devise chemical systems for incorporation of these isotopes into therapeutic radiopharmaceuticals capable of selectively delivering the isotope *in vivo* to the target tissue.

Prominent among the elements that provide a selection of radioisotopes with therapeutically useful radiological properties is rhenium. It is attractive for a variety of reasons:

- There are two β-emitting isotopes, ^{186}Re and ^{188}Re, that have potentially useful therapeutic properties. Their β-particles have energies that give a suitable range in tissues, and their half-lives are within a range that is appropriate for human use, while not presenting a long-term environmental radiation hazard. Their nuclear properties are summarized in Table 1.
- They can be manufactured economically on a scale at which widespread clinical application is feasible. The manufacturing processes are discussed further below.

Table 1. Nuclear Properties of Rhenium Radionuclides

	Half-Life (h)	Max. β- Energy (MeV)	Avg. β- Energy (MeV)	Range in Tissue (mm)	γ-Energy (KeV)	Production
Re-186	90	1.07 (71%) 0.93 (21%)	0.36	5	137 (9%)	$^{185}Re + n$ (reactor)
Re-188	17	2.1	0.77	11	155 (15%)	^{188}W decay (generator)

- The chemical analogy with technetium means that much of the research and development that has culminated in the range of technetium-based imaging agents available today, will be directly transferrable to therapeutic analogues. As with technetium, the variety of kinetically stable complexes accessible encompasses a wide range of ligand types, and potentially a wide range of metal oxidation states (+1 to +7).
- Under biological conditions, the "thermodynamic end-point" (the ultimately most stable form of the element) is perrhenate, which is rapidly excreted by the kidneys. This is in contrast to, for example, ^{131}I and ^{90}Y, which thermodynamically tend towards, respectively I^- (which accumulates in thyroid) and solvated Y^{3+} (which accumulates in skeleton); each of these constitutes a radiological hazard to nontarget tissues.

The application of rhenium radioisotopes in treatment of cancer (and of other diseases) has now reached a relatively advanced stage: some agents, such as the ^{186}Re-diphosphonates, have recently become commercially available as licensed drugs, and others are in various stages of development ranging from preliminary chemistry to clinical trials. This review summarizes the present status of rhenium radiopharmaceuticals. It is intended to be reasonably exhaustive, covering all literature concerning nuclear medicine-related applied chemistry and radiochemistry of the radioisotopes, together with the basic inorganic chemistry where directly relevant. Patents are covered only where they provide actual experimental examples.

2. PHYSICAL, CHEMICAL, AND BIOLOGICAL BACKGROUND

2.1 Nuclear Properties and Radiation Dosimetry

When selecting radionuclides for targeted radiotherapy, there are a number of factors to take into account. As well as having chemical properties that permit stable attachment to a suitable targeting principle or carrier, the radionuclide must have appropriate nuclear properties. Most important is the type of emission which for therapy must be particulate nuclear radiation (α, β) or suitable secondary electrons (electrons displaced from the atom rather than the nucleus, as a consequence of the nuclear disintegration) such as Auger electrons. The particles emitted must have an energy that gives an appropriate range in tissue. This may vary from less than 1 μm to several mm. Since the prime radiosensitive target in cells is the cell nucleus, the very low-energy (range less than 1 μm) β- or Auger-emitters must be delivered directly into the nucleus in order to impart nuclear damage. If this can be done, the resulting cytotoxicity is very high. Higher energy emitters need not be delivered to the nucleus, or even necessarily into the target cells, since their emissions can travel many cell diameters, and indeed impart greater damage towards the end of their path. However, the damage they inflict is more indiscriminate and can result in toxic doses to normal tissues surrounding the tumor. Also, the molecular binding site or receptor to which the tracer is targeted may be nonuniformly distributed within a tumor, and not all malignant cells may possess them. In such cases the "cross-fire" from the higher energy emitters is an advantage. Thus, low-energy emitters are particularly appropriate when the target is single cells or small clusters of cells that have yet to grow into an identifiable tumor, whereas higher-energy emitters are more useful against large tumor masses. It is also of value if the radionuclide emits γ-photons or positrons, so that the localization of the tracers can be determined by gamma or positron imaging. These emissions should be of low abundance (e.g. 10% of disintegrations) otherwise they will contribute significantly to the dose to nontarget tissue and to other individuals (family and nursing staff).

A second important parameter is the decay half-life of the radionuclide. It must be long enough to give sufficient time for pharmaceutical preparation and *in vivo* accumulation in the target, but not so long as to impose a prolonged hazard to other personnel or cause an environmental problem. A few days is generally most appropriate, especially if the target localization process takes 24–48 h (as in the case of radiolabeled antibodies for example, which localize slowly because of their high molecular weight—typically ca. 150,000). For more rapid targeting processes (e.g. with small peptides or colloidal agents), a half-life of 12 h or more may be sufficient.

^{186}Re and ^{188}Re offer suitable properties in many of these respects, and ^{186}Re was identified early as a preferred radionuclide for targeted radiotherapy [1] on the basis of both its nuclear properties and its chemistry, which is suited to stable bioconjugate formation. Although no experimental comparison of the lethality of the two radionuclides in tissue has been reported, theoretically each has its own advantages and disadvantages. The half-life of ^{186}Re (about 4 days) is well suited to delivery with labeled antibodies, while ^{188}Re has a shorter half-life (17 h) which is at the lower limit of that which is considered useful for antibody-mediated delivery. However, delivery with antibody fragments, peptides and other small molecular forms is quite feasible. Because of the shorter half-life of ^{188}Re it is expected that, given optimal tumor size (see below), about 20% more ^{188}Re radioactivity must be taken up per gram of tumor than ^{186}Re [2].

^{186}Re has a moderate β-energy (max 1.08 MeV) giving a range of about 5 mm in tissue, appropriate for macroscopic tumors without excessive irradiation of surrounding non-target tissue. The β-energy of ^{188}Re is high (max 2.1 MeV) giving a higher total dose delivered per disintegration event and a range in tissue of about 11 mm, which is highly inappropriate for small tumors but potentially very effective in large masses. Both radionuclides also give rise to secondary electron emissions (Auger and conversion electrons) and these add to the radiation dose delivered within a very small radius ($\ll$ 1 cell diameter) [3]. The different range of the emissions in tissues means that target:total body dose ratio is likely to be higher for ^{186}Re than ^{188}Re [4]. Further, each radionuclide has an "optimal tumor size" for which its probability of achieving a cure is maximized: computer models indicate that for ^{188}Re this diameter is about 26 mm while for ^{186}Re it is estimated at 9 mm. On this basis, radiotherapeutic treatments involving carriers of rhenium radionuclides may well turn out to be optimal if *both* radionuclides are used simultaneously in order to maximize probability of sterilizing tumors of various sizes [2]. This possibility has not yet been investigated experimentally.

Also relevant is the mode of production. ^{186}Re is produced by neutron irradiation of ^{185}Re, and despite recent steady improvements in specific activity, the product is unavoidably diluted with nonradioactive ^{185}Re ("carrier"). This limits the specific activity of the radioisotope, which therefore requires relatively large amounts of the carrier molecule (e.g. antibody) to deliver adequate quantities to the target. In some circumstances the large amount of carrier molecule may elicit an undesirable physiological side effect (e.g. an immune reaction) or saturate the binding sites. Specific activities of >150 GBq mg^{-1} are now routinely available, permitting therapeutic doses (e.g. 7500 MBq) to be delivered with less than 50 mg of antibody. ^{188}Re on the other hand is most conveniently and economically produced by decay of ^{188}W. Since rhenium and tungsten compounds are chemically separable, the ^{188}Re is free of nonradioactive rhenium (i.e. is essentially "carrier-free") and has much higher specific activity. A dose of 7500 MBq ^{188}Re contains about 10^{-9} moles of rhenium, so that a therapeutic dose can in principle be delivered with much less than 1 mg of antibody.

2.2 Biological Behavior of Rhenium

Rhenium has never been shown to be an essential trace element, although in trace quantities perrhenate has been seen to enhance population growth in the single-celled eukaryotes *Tetrahymena pyriformis* and *Chlamydomonas reinhardtii* [5]. Nevertheless its behavior in biological systems has to be understood as a background to the behavior of rhenium-based radiopharmaceuticals. As mentioned above, because of the relative ease of oxidation of rhenium to its highest oxidation state, the thermodynamic end point under aerobic biological conditions is perrhenate. The similar size, charge, and pseudo-spherical symmetry of the perrhenate ion and the iodide ion results in perrhenate acting as a substrate for the iodide transporter present in thyroid. Perrhenate is thus accumulated in thyroid (although to a lesser extent than pertechnetate [6]), but unlike with iodide there is no trapping mechanism so the uptake is relatively transient and indeed can be inhibited by administration of perchlorate [6a]. This is particularly relevant to the risk of rupture of Re-188-filled angioplasty balloons inserted into coronary arteries to prevent restenosis [6a]. Significant doses of perrhenate, like perchlorate [7] have been shown to inhibit uptake of iodide in the thyroid. Perrhenate is rapidly excreted via the kidneys by glomerular filtration, more rapidly than pertechnetate [6]. Rhenium in the form of sodium perrhenate has very low toxicity: the lowest dose to cause death is 1320 mg/kg [8]. Since rhenium in oxidation states below VII has an affinity for sulfur ligands, other possible pathways followed by rhenium could include binding to thiol-rich proteins such as metallothionein. There is evidence that Re(V) binds to metallothionein *in vitro* [9].

2.3 Rhenium Chemistry

This review is an inappropriate medium for a detailed discussion of rhenium coordination chemistry, but a brief summary of the trends is required as a background against which to set radiopharmaceutical developments, and is given below.

All formal oxidation states of rhenium between (-I) and (VII) are represented by known compounds. The most stable and readily accessible are (VII), (V), and (I), although the intermediate oxidation states are by no means rare. The higher oxidation states are typically stabilized by highly electronegative strong π-donor ligands (O^{2-}, NR^{2-}, N^{3-}, F^-) as in, for example, ReO_4^- (the most stable form of rhenium under aqueous, nonreducing conditions). Tetrathiorhenate(VII), ReS_4^-, is known but is readily hydrolyzed in the absence of a source of excess sulfide.

As the formal oxidation state is lowered, the need for stabilization by π donors is diminished. Thus although rhenium(V) chemistry is dominated by oxo-complexes (with a few notable exceptions (e.g. $ReCl_5$, Re–hexathiolate complex [10]), rhenium(V) complexes have only one or two terminal oxo-ligands compared to three or four in the case of rhenium(VII). Indeed, even within the oxidation state (V), the number of oxo-groups required can be controlled by the π-donor ability of the ancillary ligands: π-donating alkoxide and thiolate ligands stabilize the ReO^{3+} core, while the non-π-donating amines stabilize the ReO_2^+ core. Further lowering of oxidation state removes the need for oxo- or nitrido-ligands altogether: halides and pseudohalides are sufficient, and π-acceptor ligands (e.g. tertiary phosphines, carbonyls) become relatively more important. Rhenium(I) compounds without strong π-accepting ligands (e.g. carbonyl, arene, cyclopentadienyl) are virtually unknown.

The full exploitation of the wide range of oxidation states of rhenium in radiopharmaceutical design requires not only a wide range of ligand donor types with varied π-donor and acceptor properties, but also suitable reducing agents to provide access to oxidation levels below (VII). This is because for radiopharmaceutical purposes the most conveniently manfactured form of rhenium radionuclides is perrhenate. The traditional routes to reduced rhenium complexes are not particularly amenable to use in a radiopharmaceutical context, where mild, rapid, and preferably aqueous routes are most desirable. For example, a large fraction of the known coordination compounds of rhenium are prepared via the complex $ReOCl_3(PPh_3)_2$, which is itself prepared from perrhenate by reduction with phosphine in boiling ethanol/concentrated hydrochloric acid [11]. Suitable mild aqueous routes are, in fact, already available for rhenium(V). The most useful reducing agents for direct reduction of radioactive perrhenate have been found to be stannous chloride and chelates of stannous ion. Complexes in still lower oxidation states, however, commonly require even harsher conditions (powerful reducing agents such as metal hydrides, high concentrations of π-acceptor ligands such as carbon monoxide, requiring use of high-pressure equipment, and high temperature). Thus, the most recent significant chemical development in the field has been the devising of mild routes to highly reduced rhenium complexes (oxidation states I–III; see Section 4.1).

Rhenium occupies group VII of the periodic table, and its chemistry closely mimics that of technetium except in those aspects that differ characteristically between second- and third-row transition elements: redox potentials and ligand exchange kinetics. Thus, although rhenium complexes and the corresponding technetium complexes are usually analogous in structure (see ref. 6 for examples), rhenium compounds have lower redox potentials and are less reactive. In pairs of analogous complexes that have been directly compared, the technetium analogue is always easier to reduce and harder to oxidize. The quantitative difference in redox potential within the pairs is variable, however. A summary of the available data over a range of oxidation states is provided by Deutsch and co-workers [6].

Rhenium–ligand bonds are predictably less labile than technetium–ligand bonds in reactions that proceed via simple associative or dissociative mechanisms. For example, in $[MO_2py_4]^+$ the first order rate constant for pyridine dissociation is ca. 8000 times higher for Tc than for Re [12,13]. However, protic ligands bound to Re will have higher pK_a's than when bound to Tc (due to the same greater "electron richness" that gives rise to lowered redox potentials and enhanced π-back-bonding compared to technetium) [14]. This could lead to acid-catalyzed ligand dissociation (e.g. of thiolate ligands) being more rapid, under a given set of conditions, for Re than for Tc.

The periodic trends that distinguish the chemical behavior of rhenium and technetium are manifested in differences between the biological behavior between analogous rhenium and technetium compounds. The comparison can shed light on the mechanism of accumulation in target tissues. An example of the effect of difference in redox potential is provided by the behavior of the analogous pairs of lipophilic cations $[M(DMPE)_3]^+$ (DMPE = bis(dimethylphosphino)ethane) and $[M(DMPE)_2Cl_2]^+$ (M = Tc and Re). The nonreducible M(I) $[M(DMPE)_3]^+$ species have identical biodistributions for both metals, but the M(III) species $[M(DMPE)_2Cl_2]^+$ have different biodistributions because the technetium analogue is readily reduced to the uncharged Tc(II) complex *in vivo* while the rhenium analogue is not, since it has a lower redox potential. This *in vivo* reduction led to the failure of the Tc(III) complex as a myocardial imaging agent [6]. Similar trends have been observed with the nitrido complexes $[MN(DMPE)_2Cl]^+$ [15]. The redox trends are also manifested in the greater degree to which some rhenium complexes (e.g. the diphosphonates, *vide infra*) are more readily oxidized to perrhenate than technetium analogues both *in vivo* and *in vitro*.

A greater tendency of rhenium complexes (compared to technetium analogues) to expand their coordination numbers has been invoked to rationalize the stronger interaction of the perrhenate ion with carboxylate ligands. This association has been suggested as a possible cause of the different quantitative biodistribution and excretion characteristics of pertechnetate and perrhenate: perrhenate is accumulated in thyroid to a lesser extent and renally excreted more rapidly than pertechnetate [6].

3. MANUFACTURE OF RHENIUM RADIONUCLIDES

3.1 Rhenium-186

^{186}Re is manufactured by neutron irradiation, in a nuclear reactor, of either natural rhenium or isotopically enriched ^{185}Re, in an appropriate chemical form. If natural (37.4% ^{185}Re, 62.6% ^{187}Re) is used, the product is heavily contaminated with ^{188}Re, which is removed by simply awaiting its decay. The resulting delay results in considerable waste of the desired isotope ^{186}Re through decay. This process was adequate for the supply of ^{186}Re at low specific activity, and has been used for manufacture of ^{186}Re-labeled rhenium sulfide colloid, a topical radiopharmaceutical for radiation synovectomy in arthritis (*vide infra*). However, for targeting applications where higher specific activity is desirable, enriched ^{185}Re is required as the target for neutron irradiation. The enrichment both minimizes ^{188}Re contamination and minimizes the amount of residual nonradioactive "carrier" rhenium (^{187}Re).

Various chemical forms of rhenium may be used as neutron irradiation targets. Metallic rhenium [16,17], after irradiation, may be dissolved in nitric acid or hydrogen peroxide to give solutions containing perrhenic acid [16]. These reactions are vigorous and can lead to losses due to the volatility of the product. An alternative which avoids the harsh oxidation is use of a water-soluble perrhenate salt as the target material. The counterion must be selected to minimize nuclear activation of its constituent elements and to facilitate ease of extraction and purification. The aluminium salt has been selected for this purpose [18] and much of the ^{186}Re now in clinical or developmental use is obtained by this process. The perrhenate can be concentrated and purified if necessary by precipitation from water as a tetrabutylammonium salt [19]. For research studies in which specific activity is not important, a wider selection of target materials can be utilized. For example Hafeli and co-workers [20,21] used the versatile synthesis precursor $ReOCl_3(PPh_3)_2$. After irradiation, the complex was recovered with little decomposition.

^{186}Re can also be produced by proton or deuteron bombardment of ^{186}W in a cyclotron at high or moderate energies [22]. However, the practicality of this approach for manufacture of clinical scale batches is uncertain.

3.2 Rhenium-188

The most useful route to ^{188}Re is by decay of ^{188}W. The half-life of ^{188}W is approximately 60 days, and this parent/daughter relationship is amenable to construction of a satisfactory [23] generator system, analogous to that used daily in hospital radiopharmacies worldwide for supplying technetium-99m [24]. The core of the generator is a solid phase (usually alumina [24–29]) to which is chemisorbed ^{188}W-tungstate, produced by irradiation of ^{186}W-enriched tungsten oxide. The target material is dissolved in 0.1 N NaOH or 0.1 N NaOH/sodium hypochlorite [30] and loaded onto the generator column. The decay product, ^{188}Re, is readily eluted with 0.9% (physiological) saline as sodium perrhenate with no of ^{188}W; any trace of ^{188}W tungstate is readily tested for by a simple Sep-Pak® filtration step [31].

The manufacture of the parent radionuclide ^{188}W requires a *double* neutron-capture reaction of ^{186}W, and as a result is relatively inefficient. A high neutron-flux reactor is required, and the irradiation requires long periods at high neutron flux (e.g. 3×10^{14} neutrons/cm^2 s^{-1} for 7 days) [18,32]. Even then, the specific activity of the ^{188}W is low (by comparison to the ^{99}Mo used for ^{99m}Tc generators) and the product contains large amounts of inactive ^{186}W carrier. This requires a relatively large solid support in the generator column, which in turn requires elution with a relatively large volume (at least 10 mL) of eluent (saline).

To overcome this problem, modifications to the design of the basic ^{99m}Tc generator (based on an alumina solid support) have been devised. In the most effective and simple method, the volume of the eluate can be reduced from >10 mL to <2 mL by a simple sample preparation column step (Figure 1): first the eluate is passed through a small disposable cation exchange column loaded with Ag^+ ion (AG, Dionex UK) to remove chloride ion, then immediately through a strong anion exchange column (e.g. Analytichem SAX, Jones, UK) which binds all perrhenate. The radioactivity can be eluted quantitatively from this column in 2 mL 0.9% saline or 1 mL of more concentrated saline, for use in radiopharmaceutical synthesis [33,34]. In a related method (the "tandem generator") the column is eluted with ammonium nitrate directly onto an anion exchange resin, which is then washed with water and dilute nitric acid, and the ^{188}Re then eluted as perrhenate/perrhenic acid with 1.8 M HNO_3 (10 mL) [35–38]. An alternative approach is to remove the need for a bulky solid support by precipitating the ^{188}W tungstate as an insoluble zirconium tungstate [18,39–43] which is packed into an elutable column, from

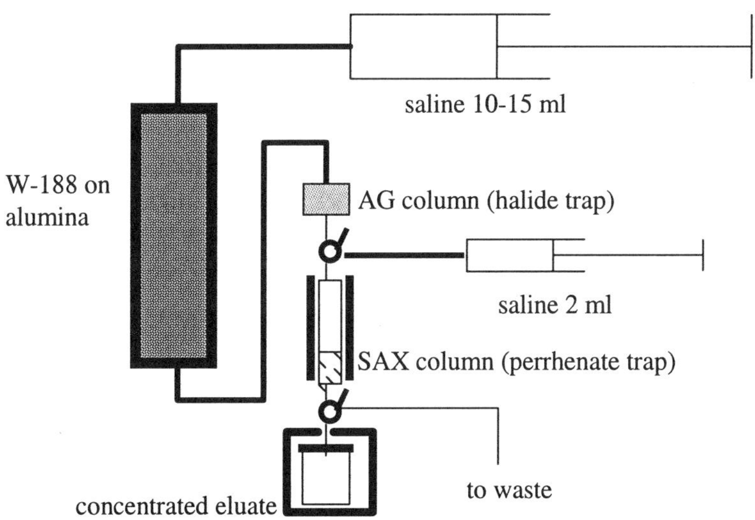

Figure 1. Rhenium-188 generator and disposable column system for concentrating eluate to small volumes.

which ^{188}Re-perrhenate is eluted with a relatively small volume of saline (e.g. 5 mL for a therapeutic scale generator). A similar approach [32] employs insoluble aluminium tungstate as the irradiated target and column packing, and the ^{188}Re is eluted as aluminium perrhenate (15–20% yield) with saline in a volume of 5–10 mL.

Another problem with generators of therapeutic scale is autoradiolysis due to the high activity of high-energy β-emitter, especially when the generator is stored "wet" (i.e. without removing residual eluent). This reduces elution yields to below 50%. The loss can be offset either by storing the generator dry or by including 0.01% ascorbic acid in the eluant [44].

The inefficiency of the double-neutron capture process, and the resulting requirement for unusually high-flux reactors, places limitations on the scale of potential use of ^{188}Re. However, this may be offset by the major advantages that derive from the availability of a generator. A single generator containing, say, 0.5 Ci of ^{188}W (manufacturing costs ca. $4000) can potentially supply therapeutic doses of ^{188}Re to several hundred patients during its lifetime of 2–6 months. The doses would be available daily at short notice, and in principle a regional radiopharmacy in possession of a generator could supply doses to several hospitals within a few hours radius. Thus, ^{188}Re has the potential to become a very cost-effective radioisotope for therapeutic applications [44a].

4. BIOCONJUGATE CHEMISTRY OF RHENIUM

One of the most important classes of rhenium-containing radiopharmaceuticals for selective delivery is the bioconjugate type: biological molecules such as antibodies and peptides tagged with a rhenium radiolabel. Many reviews and articles have appeared on the subject of labeling antibodies with metal ions, among which rhenium figures prominently [45–47]. There are two fundamental approaches to bioconjugate synthesis: attaching rhenium by means of a synthetic *bifunctional chelator* covalently attached to the biomolecule, and attaching the rhenium to a native metal-binding site in the molecule (*direct labeling*).

4.1 Bifunctional Chelators

The purpose of a bifunctional chelator is to link a radiometal covalently to a targeting molecule such as a monoclonal antibody or a receptor-binding peptide. In the case of rhenium, the molecule therefore contains a moiety for chelating rhenium with (ideally) the greatest attainable kinetic stability under biological conditions, linked to a reactive group capable of forming a covalent bond with the antibody or peptide. This group is usually selected to react with primary amine groups present as lysine residues or the amino-terminus, or less often, to other groups such as thiols (which may be present naturally or following covalent modification of lysine residues with compounds such as Traut's reagent, 2-iminothiolane), or antibody carbohydrate groups (which can be modified to provide a reactive aldehyde group). Many such reactive groups have been incorporated into bifunctional chelators: activated esters (e.g. tetrafluorophenol esters, *N*-hydroxysuccinimide esters, imidate esters), isothiocyanates, carboxylic anhydrides, vinyl groups, and others. The selection of these groups is incidental to this review, and in the following discussion the material is organized according to the type of chelating groups employed.

Bifunctional chelators must be capable of conjugation under mild, aqueous conditions reasonably close to physiological pH, in a relatively short time, with minimal need for purification procedures. For greatest convenience, it is most appropriate first to conjugate the chelator to the protein to form a bioconjugate that can be stored, then add the radiolabel ("postformed" labeling approach). However, the relative kinetic inertness of rhenium (and, consequently, the harsh conditions required for reduction and chelation or transchelation) often preclude this sequence of operations, and formation of a rhenium–chelator complex must be done first, followed by conjugation of the complex to the protein ("preformed chelate" approach). Much of the development work has aimed at either achieving effective postformed labeling, or minimizing the inconvenience of the preformed chelate method.

The earliest attempts to conjugate rhenium radionuclides to proteins were made using diethylenetriamine pentaacetic acid (DTPA, **1**) as the basis for the chelator. The cyclic bis-anhydride of DTPA (**2**) has been widely used as a very convenient bifunctional chelator for coupling indium-111 to proteins by aminolysis with lysine residues, and many workers have investigated its use for chelating other metals [48–49]. In no case, however, have the results been as satisfactory as those obtained with indium. For example, Quadri and Wessels [50] attempted labeling of protein–DTPA conjugates (**3**) with ^{186}Re–perrhenate using stannous chloride, dithionite, and hypophosphorous acid as perrhenate-reducing agents, and obtained radiochemical yields of labeled protein of up to 18%, 18%, and <0.1% respectively, along with substantial loss of antigen

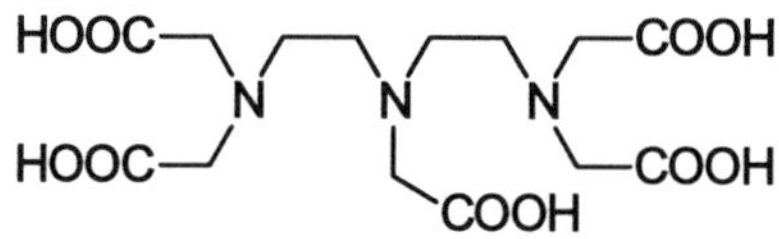

1

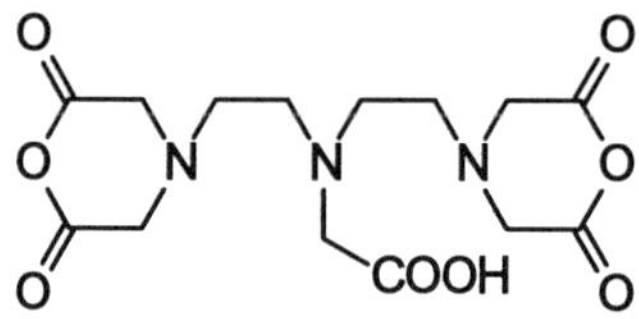

2

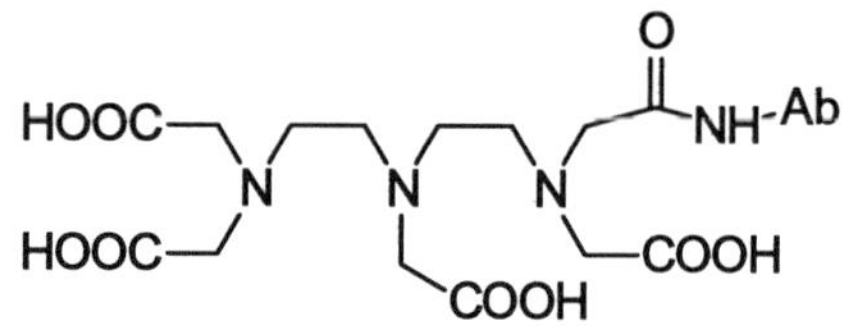

3

binding affinity and very poor *in vivo* stability. Majali [16] investigated the optimal conditions for the formation of a complex of ^{186}Re with DTPA itself using stannous chloride as reducing agent, highlighting the need for heat and low pH (in contrast to the synthesis of the ^{99m}Tc complex) to obtain satisfactory yield. The Re–DTPA complex, whose structure was not investigated, was stable in neutral pH buffer over several hours, but its stability in serum was not assessed. The need for these relatively harsh conditions, which are incompatible with labeling of sensitive biomolecules, explains the poor results in labeling the DTPA–protein conjugates. Success in labeling antibodies with ^{186}Re by means of the DTPA cyclic anhydride was achieved by using the rhenium(V) citrate complex as an intermediate for transfer of rhenium(V) to the DTPA–protein conjugate (50% yield) rather than direct reduction of perrhenate in the presence of the conjugate [51]. This approach has been used successfully with many other bifunctional chelators (*vide infra*). Despite the good yield and convenience of this method, the stability of the resulting conjugate *in sera/in vivo* has not been investigated and the technique has not been widely adopted.

The majority of more recent work on bifunctional chelators for rhenium has been more firmly based in the known coordination chemistry of rhenium. The focus has been mainly on rhenium(V), and the type of donor atom and ligand topology has been designed with a view to forming complexes of ReO^{3+} using ligands capable of forming the base of a five-coordinate square pyramidal complex with an apical oxo-group

[52], which is one of the two most common structural motifs in rhenium(V) chemistry. The other is the six-coordinate pseudo-octahedral complex containing the ReO_2^+ core with four equatorially disposed co-ligands: although ligands for chelating rhenium in this fashion have been suggested, their use in practice has been neglected. Lower oxidation states have begun to figure recently in bifunctional rhenium chelates (e.g. cyclopentadienyl Re(I) complexes).

Ligands for Binding ReO^{3+}

This class of chelators has been by far the most extensively investigated. The observation that at least some of the donor atoms must be anionic and π-donating to stabilize the five-coordinate ReO^{3+} core (based on the observed stability of the technetium and rhenium ethanedithiolate complexes $[MO(SCH_2CH_2S)_2]^-$, *vide infra*) led to variation on the theme of linear, tetradentate ligands containing a mixture of amino- or amido- and thiolato-(anionic) donors, which give complexes with overall neutral or negative charge.

Diamido–dithiolate (N2S2) ligands. The first to be introduced, initially for technetium and later for rhenium, were the bis-amido-bis thiolate ligands, of which the prototype was **4**. They form square-based pyramidal complexes of type **5**. These complexes are anionic, as both amido protons are lost on coordination of the amide nitrogens. The prototype **4** was compared with related ligands **6–9** [53] and **10–11** [54]. Kinetic stability and synthetic yield is best when the chelate rings contain five atoms rather than six [53]; they are then kinetically resistant to ligand exchange with other chelating thiolate ligands. The ligands were conveniently prepared and presented to the metal either as the free thiol, or in protected form using a benzoyl group to protect the thiol. The protecting groups were readily removed during the synthesis of the metal complexes by reduction of pertechnetate/perrhenate in the presence of excess protected ligand.

These designs found application in the synthesis of technetium complexes (e.g. the "CO_2–DADS" complex containing the ligand **12** for renal imaging) [55], and in bifunctional form, in the synthesis of bioconjugates containing technetium and rhenium. The complexes possess the necessary stability, are structurally well characterized, and are readily excreted, enhancing the clearance of any possible small metabolites of the conjugates [56]. The "structural and biological equivalence" of the rhenium and technetium complexes provides the basis of a diagnostic

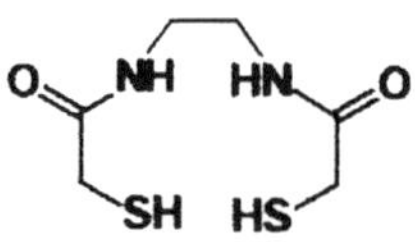

4

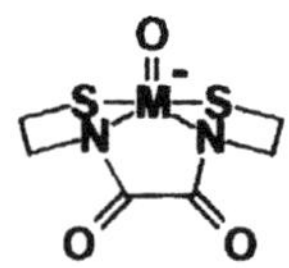

5

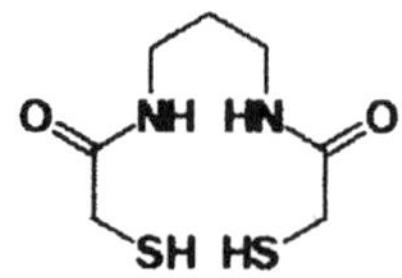

6

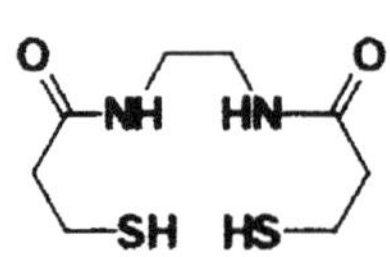

7

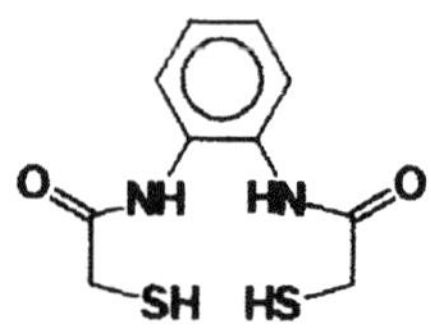

8

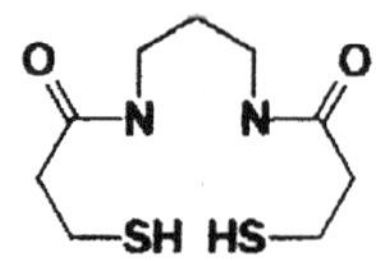

9

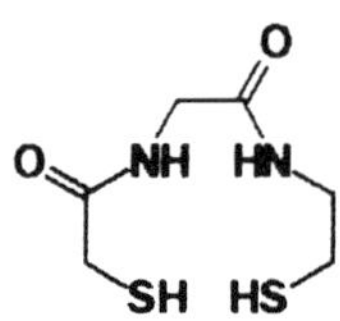

10

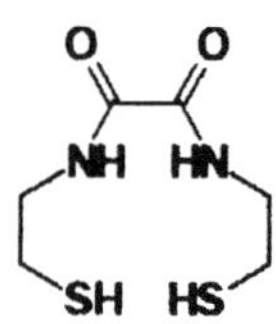

11

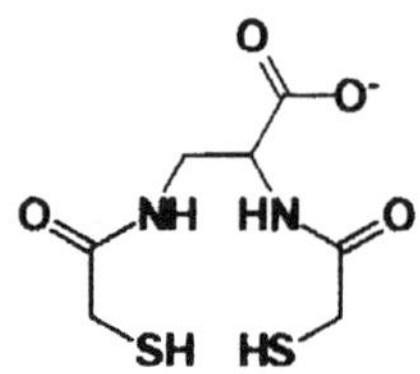

12

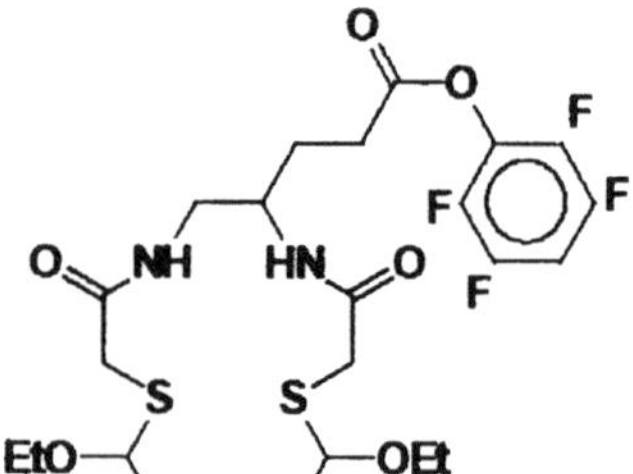

13

and therapeutic "matched pair" approach whereby the biodistribution of the therapeutic isotope is predicted by that of the diagnostic one [57]. This similarity in biological behavior of the Re and Tc complexes is surprising in light of subtle differences detected through the HPLC

syn-isomer *anti*-isomer

Figure 2. Isomers of bifunctional chelates containing the diamido–dithiolate chelating group.

behavior of the two diastereoisomers (*syn* and *anti*, designating the relative orientation of the ligand backbone substituent and the M=O bond, see Figure 2). On reversed phase (C-18) HPLC the *syn* isomers of the Tc and Re complexes elute at approximately the same time, but the *anti* isomer of the Tc complex elutes much later than that of the Re complex, despite the similarity between the structures of the corresponding isomers demonstrated by X-ray crystallography [58] and NMR spectroscopy [59]. This may reflect differences between Tc and Re in the reactivity of the vacant coordination site *trans* to the oxo-ligand.

A bifunctional ligand **13** was shown to be equally applicable to synthesis of conjugates of both technetium and rhenium, using the "preformed chelate" approach. It was designed by a series of modifications of **12**: the thiol protecting groups were changed to 1-ethoxyethyl to give improved solubility and ready deprotection during complexation; the carboxyl group was esterified with 2,4,5,6-tetrafluorophenol to provide an amine-reactive group that is stable under the acidic conditions of the complexation reaction but reactive under mild alkaline conditions; and the link between the carboxylate group and the ligand backbone was extended to prevent intramolecular reaction of the active ester. The use of this linker to prepare bioconjugates [56] is shown in Scheme 1. The rhenium chelate was formed at elevated temperature under acid conditions, taking advantage of the hydrolytic stability of the active ester and the ease of reduction of perrhenate under these conditions. Then at room temperature the pH was adjusted to 9–10 and antibody was added at a molar ratio of 3 chelates per antibody. At this pH, but not at pH 7 [60], the active ester is reactive to aminolysis (and hydrolysis) and the chelate becomes conjugated to antibody lysine residues. The overall yields were around 30–50%, and the conjugate was stable in serum over 3 days at 37

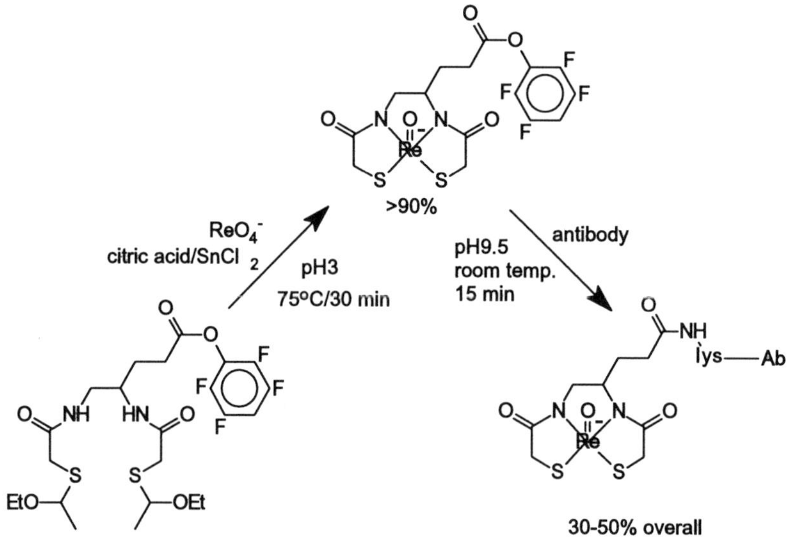

Scheme 1. Preparation of rhenium bioconjugates using the diamido–dithiolate bifunctional chelator.

°C, although about 10% decomposition per day to perrhenate was observed in phosphate buffer. At a final chelate:antibody ratio in the conjugate of 1.2, no significant loss in antibody immunoreactivity was observed. The biodistribution was identical to the similarly labeled ^{99m}Tc–antibody conjugate over 24 h in mice [57].

Diamino–dithiolate ("DADS") ligands. Diamino–dithiolate ligands of the type **14** have been used to form uncharged Tc complexes used as regional cerebral blood flow imaging agents by virtue of their ability to cross the blood–brain barrier. The complexes are approximately square pyramidal, and are uncharged as a result of loss of one amino proton from a coordinated amino group. The pK_a of this N–H bond in the complexes of the analogous non-esterified ligand **15** is 6.8 for Tc and 7.2 for Re [14]. Thus both complexes have an uncharged core at mildly alkaline pH. The difference in pK_a between Re and Tc reflects the higher electron richness and hence greater basicity of the rhenium complex. The related, more lipophilic ligand **16** similarly forms an uncharged square pyramidal complex with the ReO^{3+} core; under the synthetic conditions employed, the *syn* isomer (designating the orienta-

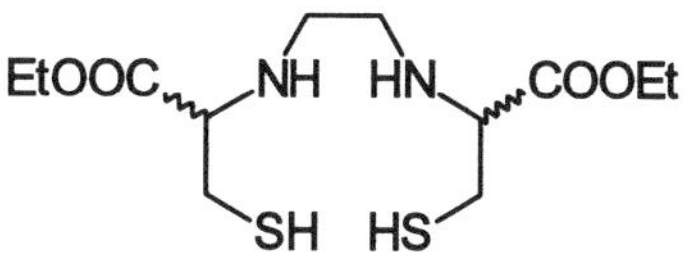

14

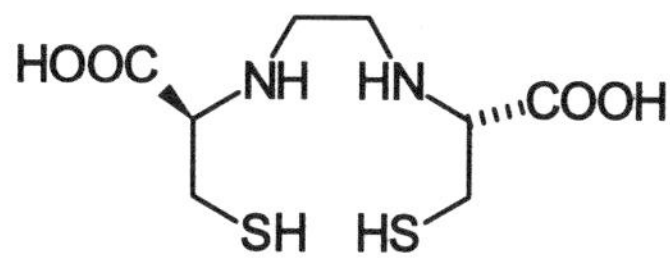

15

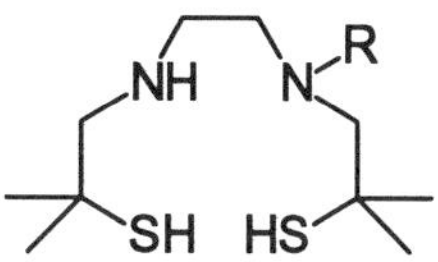

16
R = Me or alkyl chain

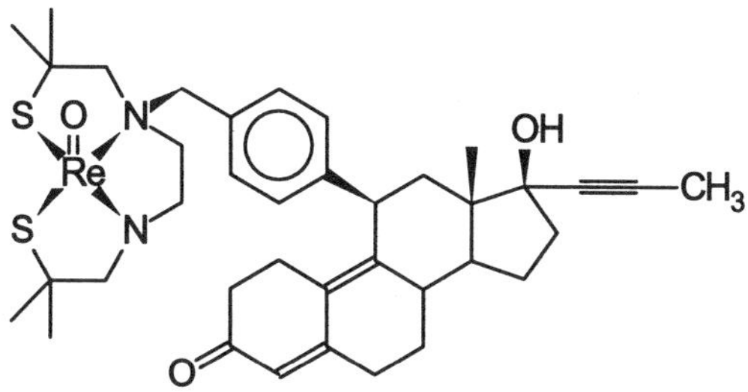

17

tion of the N-methyl group with respect to the Re=O bond) is strongly preferred [61]. The uncharged, lipophilic nature of the rhenium complexes of ligands **16** has been exploited to solubilize radioactive rhenium in lipiodol (iodized poppy seed oil), which when administered intra-arterially to the liver is selectively accumulated in hepatocellular carcinoma, thus providing a potential targeted radiotherapy treatment for this disease [62].

This ligand system has also been used to form receptor-targeting conjugates with progestins [63,64] with structure **17**. The position at which the chelator is attached is the only one of four investigated that gave retention of receptor binding. The specific diastereomer shown in **17** had the highest receptor affinity of the four stereoisomers. The conjugate was prepared by direct stannous reduction of perrhenate under acidic conditions at room temperature in the presence of the ligand. Despite a high degree of nonspecific binding due to its high lipophilicity, it showed good uptake in the target tissue (uterus) in rats with promise that modification to the chelate linker chemistry will reduce nontarget uptake to a satisfactory level for therapeutic application in receptor-positive tumors.

Despite the mild chelating conditions, there are no reports of the diamino–dithiolate ligands being used to label antibodies, perhaps because their high lipophilicity prevents use in aqueous media.

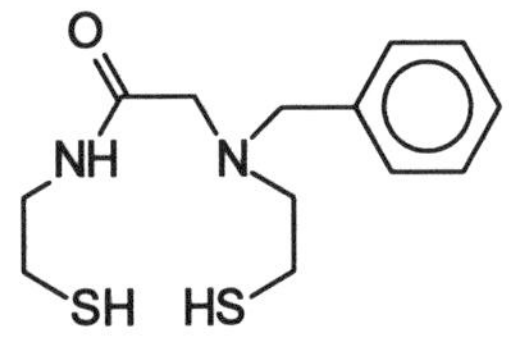

18

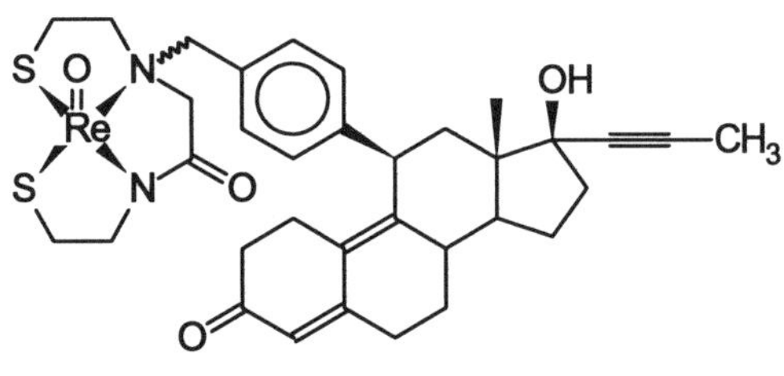

19

Monoamino–monoamido–dithiolate ("MAMA") ligands. In an attempt to reduce the lipophilicity of the steroid adduct **17**, the chelator was redesigned to include both an amino- and an amido-donor group (**18**) in order to preserve the charge neutrality but increase polarity. The X-ray crystal structures of the *syn* and *endo* isomers of the rhenium complex of **18** show that the usual distorted square pyramidal structure is maintained, and the complex was uncharged as expected [65]. The ligand system was incorporated into the steroid conjugate **19**. The results were encouraging in that lipophilicity and nonspecific binding was substantially reduced, without loss of affinity for the target receptors.

Triamido–monothiolate (N3S) ligands. The N3S ligand type was first introduced as the mercaptoacetyltriglycine (MAG_3, **20**) complex of technetium [66], a complex that is very rapidly excreted renally and has become established as a renal function imaging agent. This ligand type forms five-coordinate square pyramidal complexes with the ReO^{3+} group that have been characterized crystallographically [67,68]. All four protons (one thiol and three amide) are lost upon coordination, resulting in a monoanionic core.

Application of this ligand system for conjugation of ^{186}Re to antibodies (NR-LU-10, a murine antibody to a membrane glycoprotein associated with adenocarcinomas) has been reported in animals and humans. The conjugation chemistry is similar to that used for the N2S2 system above (Scheme 1), and again the side-chain length was selected to minimize heterocycle formation during synthesis of the active ester [69]. The active ester **21**, with the thiol suitably protected, is heated with acidic perrhenate and stannous ion and citrate at 75–90 °C, and the chelate so formed is then treated with antibody at pH 9.5 for 15 min to give 30–40% overall radiochemical yield after purification. The conjugate was stable to DTPA and serum [70–72]. This ^{186}Re conjugate with NR-LU-10 and one similarly prepared with the $F(ab')_2$ fragment of NR-CO-02 have provided encouraging results in animals [73] and in Phase I clinical trials in patients with various epithelial carcinomas [74–77] and ovarian carcinoma [78–79]. A mouse–human chimaeric antibody NR-LU-13, labeled with ^{186}Re by the same method, has also been tested in humans and shown to possess reduced immunogenicity compared to the NR-LU-10 [80].

Further developments with this system include incorporation into an avidin–biotin targeting protocol using the biotin–rhenium conjugate **22** to increase selectivity for target tissue [81]. A modified bifunctional chelator, incorporating the same chelator design but a metabolizable linker (tetrafluorophenyl mercaptoacetyl-glycyl-glycyl-seryl-succinate) has been introduced with the aim of accelerating excretion of nontarget bound radioactivity. Used in conjunction with the antibody NR-LU-10, this modification led to increased renal excretion of radioactivity while maintaining tumor targeting, thus improving the target:nontarget dose ratio [82].

Gerretsen and co-workers used a modified methodology to incorporate ^{186}Re into antibodies via the MAG_3 chelator, activating the ester *after* chelating the rhenium [83]. In this method the perrhenate was reduced directly with stannous chloride (no citrate) in the presence of the acetyl-protected MAG_3 ligand **20** at alkaline pH. Best results were obtained in this step when the solution was boiled to dryness and the dry residue heated for a further 15 min. This complex was esterified with tetrafluorophenol in the presence of a carbodiimide at mildly acidic pH, and the active ester, after purification, conjugated to the antibody at pH 9.5. The overall radiochemical yield from this rather cumbersome procedure was 55%. Most of the losses resulted from hydrolysis of the active ester in the final labeling step [84].

A similar chelator design but with an alternative antibody coupling chemistry was developed by Ram and Buchsbaum [85]. This bifunctional chelator (**23**) contains an isothiocyanato group which reacts with antibody lysine amine groups to form a thiourea linkage. The chelation (via the citrate intermediate at high temperature) was performed prior to conjugation with antibody at pH 9.2, i.e. conditions similar to those used with the tetrafluorophenol active ester. The overall radiochemical yield was 13%. A further modification, introducing the *p*-nitrophenyl ester of **20** as the bifunctional chelator, was used [86] to label the chemotherapeutic drug daunorubicin via its amino sugar moiety, and a somatostatin analogue via coupling to the N-terminal phenylalanine of the peptide. If

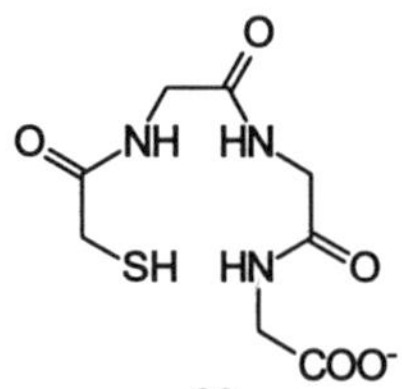

20

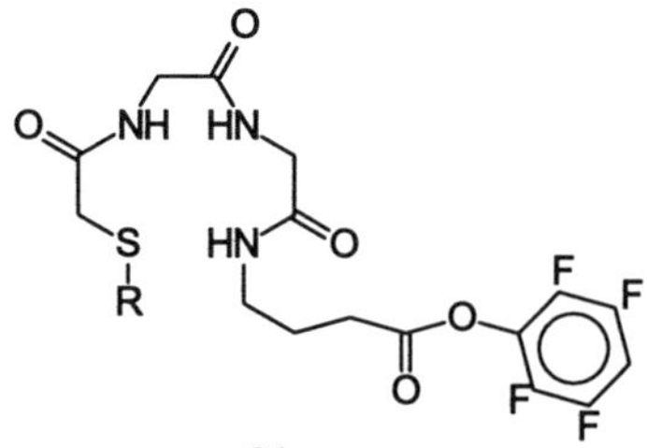

21

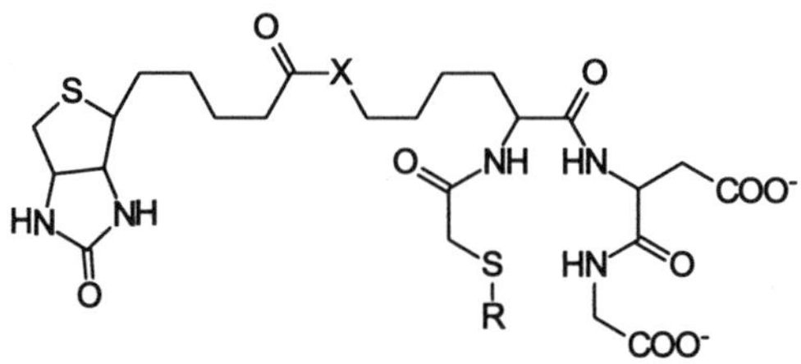

22

X = NH or $NHCO(CH_2)_4NH$

R = tetrahydropyranyl protecting group

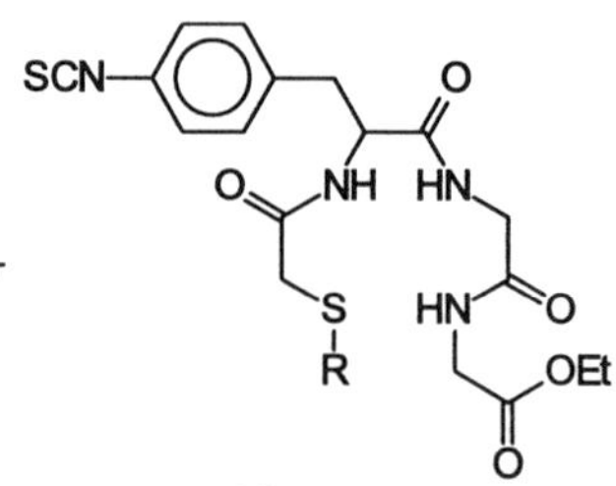

23

R = $COCH_3$

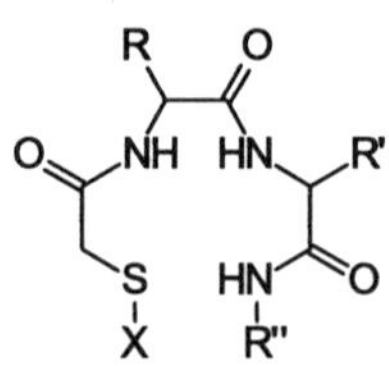

X = protecting group

R = $(CH_2)_nNH_2$

R' = $(CH_2)_2^-COOH$ or H

R" = $(CH_2)_2$-$N(CH_3)_2$ or $(CH_2)_5$-COOH

24

the chelator was first conjugated to a model targeting moiety (Boc-protected 1,6-hexane diamine as a model for a protected peptide), heating for 75 min at 90 °C was required to achieve >90% incorporation of ^{188}Re into the conjugate [87].

None of the N2S2 or N3S systems described above is suitable for the "postformed" labeling of biomolecules (i.e. coupling a chelator to the biomolecule, followed by incorporation of the radiometal into the chelator) because the unfavorable kinetics of complexation or transchelation require relatively harsh conditions (strong acid/alkali, heat) which the biomolecule would not tolerate. In order to combine the advantages of *in vivo* stability in the amido systems with the better complexation kinetics of the amine-containing ligands, new ligands have been designed containing, as well as the one thiolate and three amide groups, an amine group intended to participate in initial complex assembly process (to give an intermediate) but not in the coordination sphere of the final complex [88]. This was intended to accelerate complex formation and thus protect against nonspecific binding. There is as yet no published evidence that this approach is successful. However, there is one brief report that bioconjugates incorporating ligands **24**, containing tertiary amine substituent R, incorporate ^{186}Re from the Re–citrate intermediate in >90% yield under mild conditions [89]. Recently, the incoporation of a thioether donor into these linear, tetradentate chelators has been explored [89a,b,c].

Ligands for Binding ReO_2^+

The *trans*-dioxo complexes of general structure $ReO_2(L)_4^+$ (where L is an amino donor) are well-characterized, and when L is part of a chelating or macrocyclic ligand are kinetically stable and withstand harsh conditions (e.g. strong acid) [90]. Many workers have therefore suggested that the macrocyclic ligand 1,4,8,11-tetraazacyclotetradecane (**25**), whose trans-rhenium–dioxo complex has been crystallographically characterized [91] would be a suitable chelator for rhenium radionuclides in bioconjugates. Only recently has this ligand been evaluated as a chelator for radioactive rhenium. It was compared with the noncyclic analogues **26** and **27**. The complexes of **26** and **27** were conveniently synthesized in good yield under mild aqueous conditions ($SnCl_2$ reducing agent, 20–100 °C, pH 9–11, 2–30 min), but complexes of **25** were not: nonaqueous solvents and strongly acidic conditions were required to provide acceptable yields. Complexes of **25** and **26** showed no decomposition after incubation for 24 h in human serum, but up to 3% of the

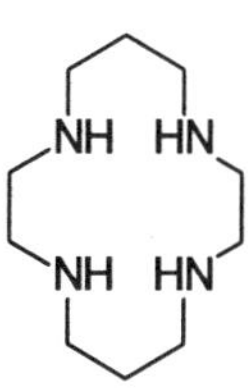
25

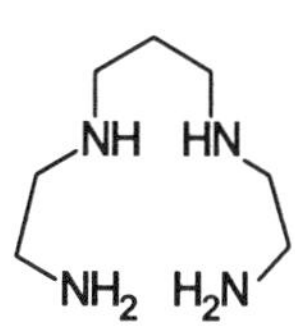
26

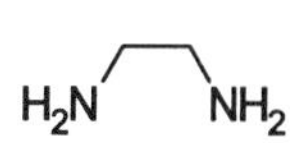
27

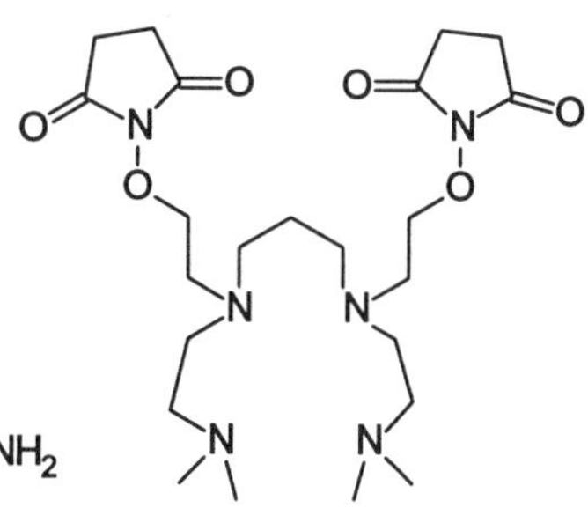
28

ethylenediamine (**27**) complex was converted to perrhenate. Thus, the noncyclic tetradentate ligand **26** offers the best combination of good stability and ease of synthesis [92]. This suggests that the tetrakis(tertiaryamine) ligand **28**, which has been partially evaluated as a bifunctional chelator for ^{99m}Tc [93], may be appropriate for rhenium too.

Bifunctional Chelators for Rhenium in Lower Oxidation States

Despite a common perception that organometallic chemistry essentially belongs in the province of catalysis rather than *in vivo* applications because of high reactivity of metal carbon bonds, certain organometallic species have very high kinetic stability. Indeed, this point is highlighted by the present wide application of the isonitrile complex "Tc(sestamibi)," [Tc(2-methoxyisobutylisonitrile)$_6$]$^+$, in myocardial imaging. Rhenium tricarbonyl and cyclopentadienyl complexes offer further ex-

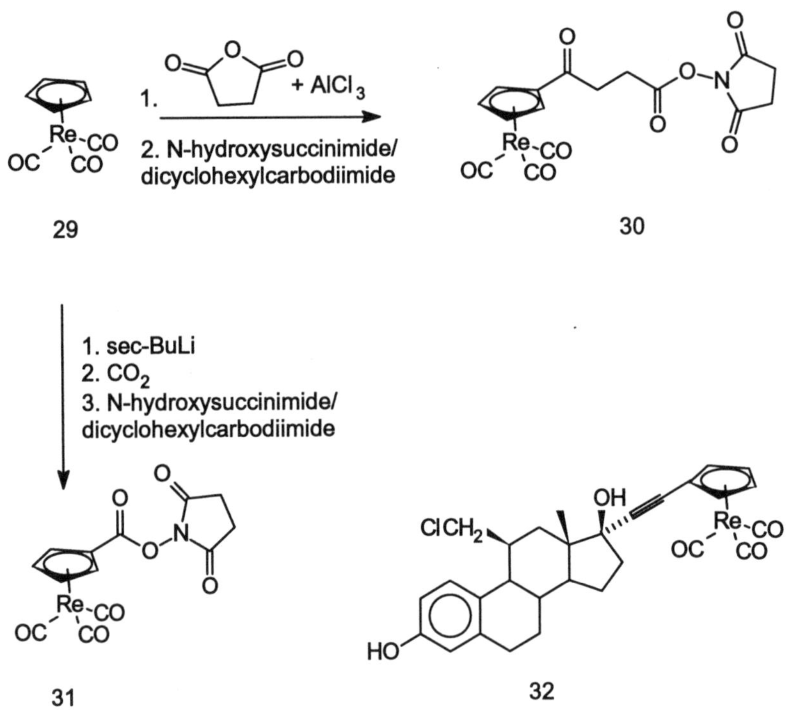

Scheme 2. Synthesis of bifunctional chelators and conjugates based on the rhenium(I)-cyclopentadienyl core.

amples of organorhenium cores with very high kinetic stability, and bioconjugates are now being developed to exploit this property. For coupling to proteins, derivatization of the Cp ligand in $CpRe(CO)_3$ (**29**) is conveniently achieved by exploiting the electrophilic substitution reactivity of the coordinated Cp (Scheme 2), to give activated ester-derivatized complexes, **30** and **31**, which were coupled to a model protein (bovine serum albumin) in good yield [94]. The steroid conjugate **32** (Scheme 2), comprising the $CpRe(CO)_3$ core coupled to an oestradiol-targeting ligand, has been synthesized and shows high affinity for the target receptor [95].

The development of adequately simple and convenient methods for synthesizing complexes of this type in a short time under mild conditions has progressed dramatically recently. This apparently daunting problem appears to be quite tractable following recent advances in synthesis of low oxidation state rhenium complexes. Katzenellenbogen and co-workers [96,96a] have approached this problem by utilizing π-ligand transfer reactions of acetyl ferrocene and diacetyl ferrocene, along with a low-valent metal carbonyl complex (chromium hexacarbonyl was most effective) as a source of carbon monoxide. This gives remarkably good yields (up to 59%) of the $ReCp(CO)_3$ core directly from perrhenate in methanol at 150 °C in 1 h. Alberto and co-workers have developed (relatively) mild routes to synthesis of other rhenium(I) carbonyl complexes. Reaction of ethanolic perrhenate with dry HCl gives the Re(V) intermediate $ReOCl_4^-$, which was reduced with borane in tetrahydrofuran/diglyme followed by bubbling carbon monoxide through the solution and then heating to 115 °C for 10 h. The product, $ReCl_3(CO)_3^{2-}$, has halide ligands that are readily replaced with, for example, isonitriles and phosphines, offering a route to a variety of complexes containing the inert $Re(CO)_3^+$ core. In aqueous solution the unusual and potentially useful species $Re(CO)_3(H_2O)_3^+$ is formed [97,98,98a].

An even quicker and milder route to low oxidation state rhenium chemistry is offered by the crown thioether ligand 1,4,7-trithiacyclononane (9S3, **33**, Scheme 3) which spans oxidation states from Re(VII) to Re(I). Perrhenate reacts with 9S3 in acidified nonaqueous solvent (e.g. MeCN) to give $ReO_3(9S3)^+$ [99], which reacts further, immediately and at room temperature, with additional 9S3 and stannous ion to give the unusual Re(II) complex $Re(9S3)_2^{2+}$ [100]. Mild reduction of $Re(9S3)_2^{2+}$ with ascorbate in water induces loss of ethene from a coordinated ligand to give $Re(9S3)(SCH_2CH_2SCH_2CH_2S)^+$ (Scheme 3) [101]. The reactivity of these various complexes with other ligands has

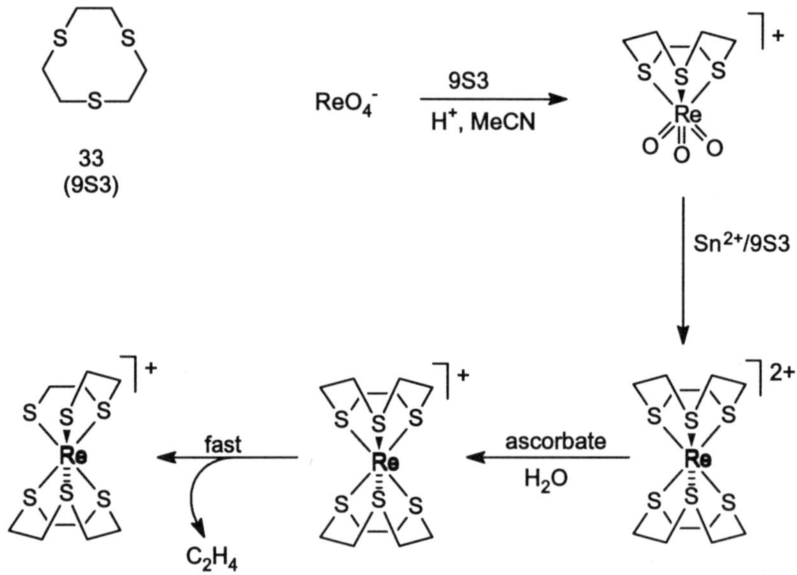

Scheme 3. Synthetic route to low oxidation state rhenium trithiacyclononane complexes.

yet to be fully investigated, but it is clear that the special coordinating properties of 9S3 offer an unusually versatile ligand environment permitting mild and convenient access to low oxidation state rhenium complexes.

Miscellaneous Indirect Labeling Approaches

Encouraging results have been reported with an N2S4 (diaminotetrathiol) type bifunctional chelator **34** (Scheme 4) whose mode of coordination to rhenium appears to be unknown, although it seems likely that a diaminodithiolate complex of ReO^{3+} is formed. A rationale has been proposed for its use in labeling antibodies via a disulfide linkage, although no evidence is presented in support of this. The conjugate is easily formed without need for "pre-chelation"—^{186}Re–perrhenate is incubated with the antibody–chelator conjugate and stannous chloride—and is stable in serum, and gives good therapeutic results in animal model tumors [102]. This system has been little studied compared to the N2S2 and N3S systems described above, but the results warrant further evaluation of the approach.

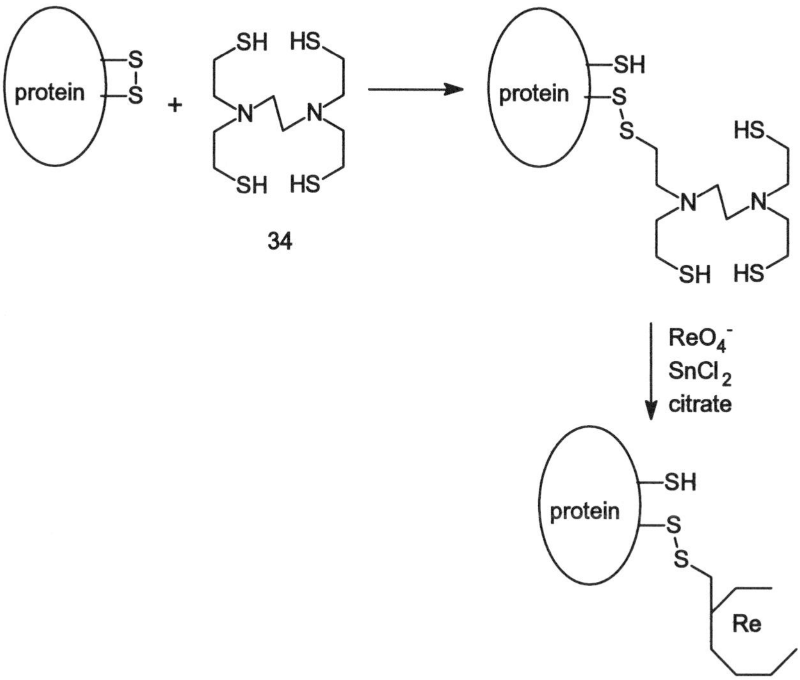

Scheme 4. Preparation of bioconjugates using an N2S4 ligand system.

A further variant on the amino–thiolate theme is the class of mono-oxo complexes with structure **35** containing one polydentate amino–thiolate ligand and one monodentate thiol. The amino–thiolate ligand is N-2-mercaptoethyl-N-(2-pyrolidin-1-ylethyl)amine, and the monodentate thiol is a thiophenol derivative such as *p*-methyl thiophenol [103,104]. This in principle presents opportunities for easy alteration of the complex properties by variation of the monodentate thiol ligand. The idea has been exploited in the design of receptor-binding rhenium complexes such as **36** in which the monodentate thiol is coupled to a series of ligands for the serotonin and other receptors [105,105a,b,c,d].

A bis(thiosemicarbazone) bifunctional chelator **37** has been used to conjugate ^{186}Re to an anti-CEA antibody (via thiol groups introduced using Traut's reagent) using the pre-formed chelate approach. The labeling proceeded in "good yield," by thiol addition across the maleimide double bond under mild conditions (pH 7.2, 1 h), without loss of immunoreactivity. No biological stability results were described, but the

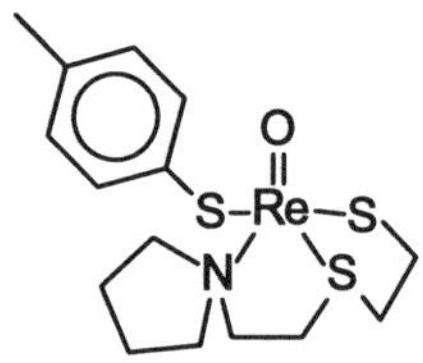

35

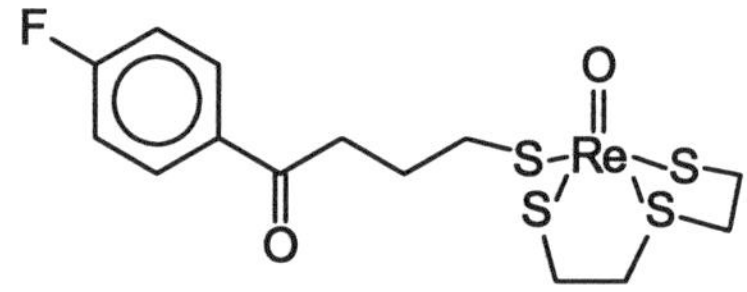

36

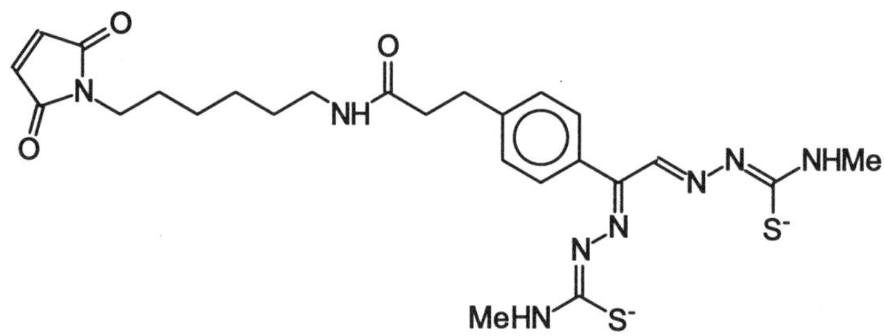

37

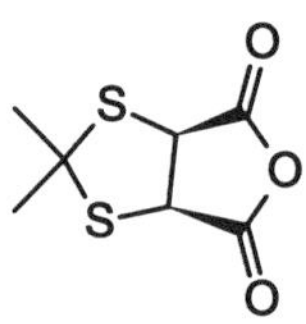

38

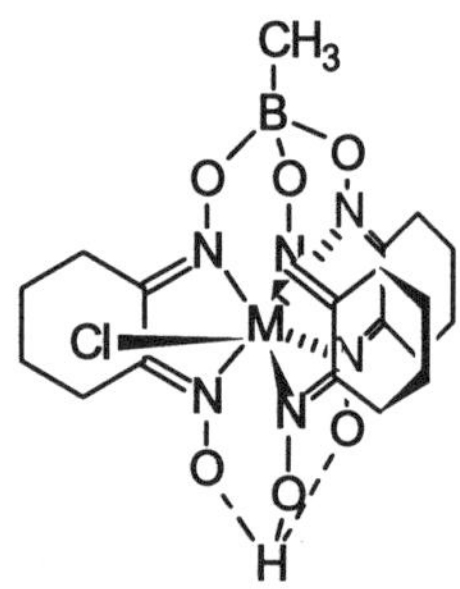

39

conjugate was reported to induce substantial reduction of tumor volume in a patient with advanced colon carcinoma after repeated treatments with therapeutic doses [106].

1,2-Dithiolate ligands form stable anionic square pyramidal complexes with the ReO^{3+} core [10]. This has been exploited in the dimercaptosuccinic acid complexes of ^{186}Re and ^{188}Re (*vide infra*), and in the use of the cyclic anhydride of dimercaptosuccinic acid as a bifunctional chelator for bioconjugate formation. The anhydride, in protected form such as **38**, is reacted with antibody or other protein to form an amide or ester linkage by reaction of lysine or serine residues with the anhydride group [107]. Radioactive perrhenate is then added together with citrate and stannous chloride to form a conjugate in which the rhenium is presumed to be coordinated by one dithiolate group and additional donors originating in the protein (e.g. amido groups).

A series of seven coordinate tris-dioxime complexes such as **39**, in which the dioxime ligands are assembled around the metal and capped with boronic acids to form a supertripodal six-coordinate ligand, have been synthesized by a "self-assembly" reaction on combination of $ReCl_3(MeCN)(PPh_3)_2$ with the appropriate dioxime and the appropriate boronic acid. The coordination sphere comprises the six nitrogen ligands of the three dioximes, and one chloride, in a capped trigonal-prismatic structure. Perrhenate or K_2ReCl_6 can be used as the rhenium-containing starting material if stannous chloride is included as a reductant. The latter approach was used to develop a kit formulation capable of synthesizing the cyclohexanedionedioxime–methylborate derivative (**39**) from ^{186}Re perrhenate in 18% yield. In principle the approach could be used to conjugate rhenium isotopes to targeting molecules by incorporating a reactive group into the alkyl substituent of the boron [108]. The chloride ligand can be exchanged for other halides, hydroxide, or thiocyanate [109].

An interesting approach to rhenium bioconjugate formation that has been developed but apparently has yet to be fully evaluated from the standpoint of applicability and stability is the use of chelating and nonchelating phosphine imine and phosphine oxide ligands [110]. For

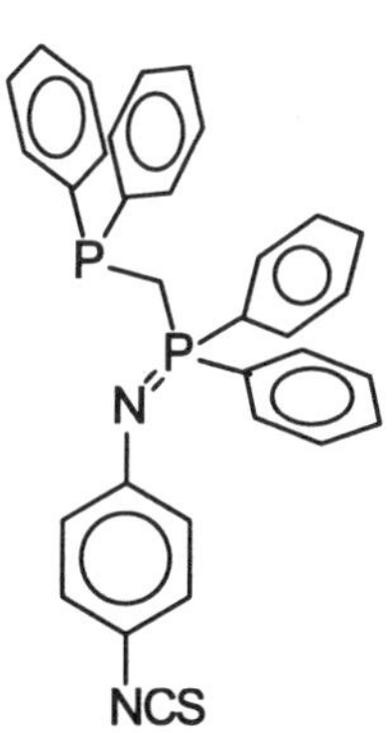

40

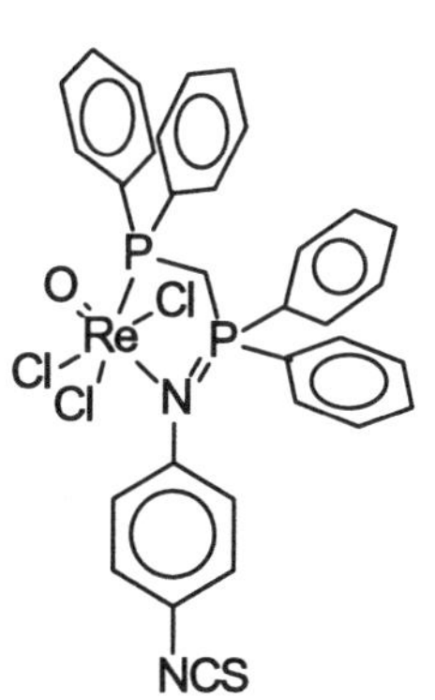

41

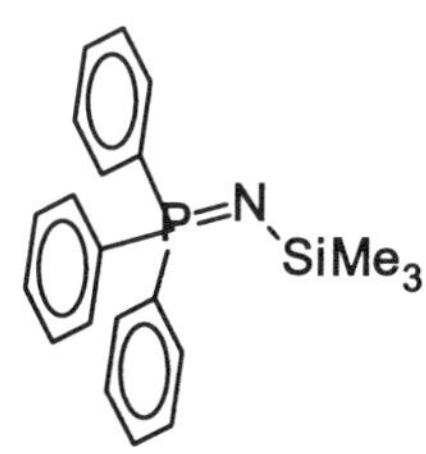

42

example, the ligand **40**, synthesized from bis(diphenylphosphino)methane and *p*-azidophenylisothiocyanate, forms a complex **41** with the $ReOCl_3$ center. This chelate was efficiently conjugated to an antibody via the isothiocyanate group. The stability properties of this conjugate are not reported. In addition, and remarkably, the trimethylsilyl phosphine imine derivative **42** reacts directly with perrhenate to give a stable Re(VII) uncharged complex. With pertechnetate the reaction occurs even in aqueous solution. This type of ligand appears to be unique to date in its ability to form a water-stable complex with Re(VII), $Ph_3P{:}NReO_3$. Whether these Re(VII) complexes are sufficiently stable to be useful in bioconjugates is not clear. However, it may be that bioconjugates formed in this way could be reduced to a lower oxidation state to form a complex of higher stability: such a process would be unique in complexing the rhenium *before* reduction of Re(VII) [111].

Finally, the use of metallothionein (a mammalian small protein in which approximately one-third of the amino acid residues are cysteine) as a bifunctional chelator has been demonstrated. Metallothionein was conjugated to an anti-carcinoembryonic antigen antibody using the coupling agent 1-ethyl-3(3-dimethylaminopropyl)-carbodiimide, then digested to $F(ab)'_2$ -metallothionein fragments with pepsin. The conjugate was combined with stannous ion and frozen for later labeling by addition of perrhenate, as described below for direct labeling of antibodies [112].

4.2 Direct Labeling

The direct labeling approach exploits functional groups already present in the biomolecule to bind directly to the radiolabel. The bioconjugates made in this way are exclusively proteins (usually antibodies and antibody fragments) and peptides, and the only functional groups capable of binding rhenium with sufficient kinetic stability are thiol groups. Usually these must be generated by deliberate reduction of native disulfide bonds. Thus, the direct method involves reduction of both the protein/peptide and the perrhenate. Different reducing agents are preferred for these two steps, so they are usually carried out separately. If the protein is an antibody, the disulfide bonds are cleaved with a suitable reducing agent (see below), purified by size-exclusion chromatography, and may then be stored in their reduced form until ready for labeling. If the protein is an antibody fragment (Fab′), the reduction step may be unnecessary because disulfide group reduction is an integral part of the antibody fragmentation chemistry. The reduced antibody can then be

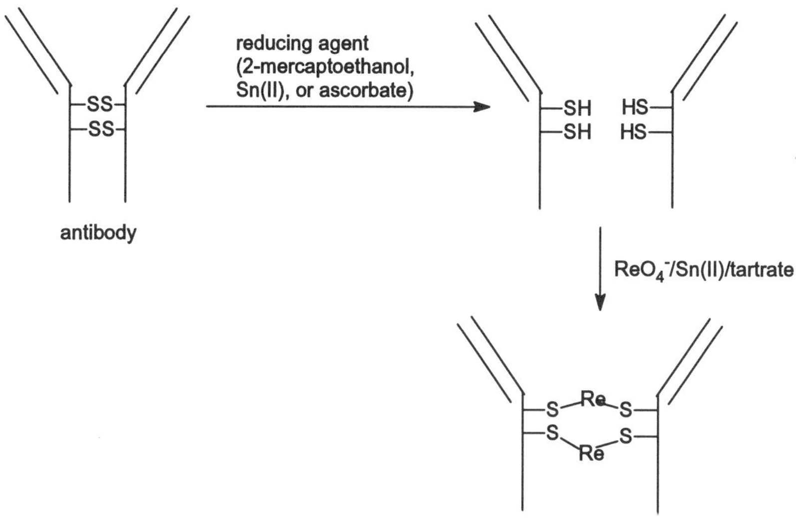

Scheme 5. Direct labeling of monoconal antibodies by disulfide bond reduction.

labeled with radioactive rhenium, which may either be prereduced and in the form of a labile complex such as the Re(V)–citrate complex, or added as perrhenate in the presence of the necessary reducing agents and transchelation agents (Scheme 5).

The antibody disulfide reduction step must be mild to prevent reduction of less accessible disulfide bonds leading to fragmentation. Both 2-mercaptoethanol [112–116,116a] and ascorbic acid [117] have been found to be effective. With 2-mercaptoethanol, the reduced antibody must be purified by size-exclusion chromatography before radiolabeling, since residual 2-mercaptoethanol can compete as a chelator for Re. Ascorbic acid has the advantage that this purification step is unnecessary: it does not complex Re, and is nontoxic. Indeed, its presence in high-activity preparations may be essential to minimize autoradiolysis. Stannous chloride is less effective than either of these reagents [117]. Recently, a photoactivation method was introduced to generate thiol groups in antibodies prior to radiolabeling with ^{99m}Tc [118]. This involves irradiation of the antibody at 300 nm to activate the antibodies, which may then be stored frozen until required for labeling. This approach has several potential advantages and evaluation of its use with rhenium isotopes is awaited.

Ascorbic acid [117], borohydride [117], dithionite [117], and stannous chloride [112–117] have all been evaluated as reducing agents for perrhenate in direct antibody labeling. They have been used in combination with methylene diphosphonate [117], tartrate [112–117], glucoheptonate [116a], and 2-hydroxyisobutyrate [117] as intermediate chelators ("transfer agents"). Only stannous chloride in combination with tartrate was satisfactory. This represents an extension of the method used for ^{99m}Tc labeling [115,119], optimized for use with rhenium by drastically increasing the amount of stannous ion and tartrate to minimize reoxidation to perrhenate during the preparation. This combination has led to the development of kit preparations, in which the 2-mercaptoethanol- or ascorbate-reduced antibody is combined with stannous ion and tartrate and the solution freeze-dried. These kits can be stored and later reconstituted with ^{186}Re- or ^{188}Re-perrhenate at 22–37 °C and pH ca. 5 for 30 min or more. The labeling yields have been optimized to the point where postlabeling purification is unnecessary (>95%) [112–117]. With ^{186}Re, because of the large excess of nonradioactive ^{185}Re, the specific activity of the labeled antibody is limited because the Re:antibody ratio does not exceed 1:1 before the labeling yield is diminished [116]. Since ^{188}Re is carrier-free, much higher specific activities can be achieved. The stability of these radiolabeled antibodies is good (<5% loss in serum in 24 h [112–117]), and the immunoreactivity is not severely impaired [112–116]. Both whole antibody and Fab′ fragments can be equally effectively incorporated into these kits for radiolabeling [112–117]. However, $F(ab')_2$ fragments are substantially cleaved to Fab′ by the high concentration of stannous ion needed to reduce the perrhenate [116].

The direct labeling approach has also been applied to small peptides containing only one disulfide bond. Varnum and co-workers synthesized a rhenium complex of the peptide RC-160, a cyclic octapeptide analogue of the peptide hormone somatostatin that binds to receptors expressed in many tumor types. The disulfide bond was reduced with ascorbate and the rhenium then introduced using the stannous/citrate/perrhenate method analogous to that described above for antibodies, at a temperature of 70 °C. A 1:1 Re:peptide adduct was formed. The structure and conformation of the adduct was investigated by high-field 2D ^{1}H NMR and molecular dynamics computations. UV spectroscopy was used to investigate the coordination around rhenium. The results suggest that the receptor-binding spatial topography of the peptide was maintained in the adduct. The UV spectra were interpreted as the result of a six-coordination environment about rhenium, with NMR shifts indicating coordination by phenylalanine, cysteine, tryptophan, and C-terminal NH_2 [120]. This analysis appears to ignore the overwhelming body of evidence that ReO^{3+} complexation with thiolate/amide/amine ligands dominates the aqueous chemistry with no tendency to form six-coordinate complexes when thiolate ligands are present. The issue of the coordination environment about rhenium in conjugates of this type needs further investigation, perhaps using EXAFS spectroscopy to identify the number and type of atoms in the coordination sphere. These studies have recently been initiated with model cysteine-containing peptides [120a].

Haberberger and co-workers also incorporated rhenium (^{188}Re) into RC-160 and another somatostatin analogue, octreotide, again via a reductive route which presumably leads to binding of Re(V) to thiolate and other donors. Stannous ion was used as reductant for both the disulfide bond and perrhenate simultaneously, with tartrate as rhenium "transfer agent," in a kit preparation. Labeling yields were >95%. Following intravenous injection in rats the conjugate was very quickly cleared from blood via the liver, possibly too quickly for intravenous administration to be viable as a therapeutic option [121]. Direct intratumor injection was therefore considered [121,121a]. For this, an alternative form of the conjugate with RC-160, prepared as a partially insoluble microparticulate form by raising the pH of the preparation, was compared with the soluble form described above. The soluble form was most effective in reducing tumor volume despite being cleared from tumor tissue more quickly, suggesting that the lack of mobility of the insoluble form prevented access to receptors resulting in a highly inhomogeneous distribution within the tumor [122].

5. "RHENIUM-ESSENTIAL" RADIOPHARMACEUTICALS

Metal-essential radiopharmaceuticals are those whose targeting properties are inherent in the structure and chemistry of the complex itself rather than a biomolecule whose targeting ability is independent of the attachment of the metal. While this class of tracers includes many well-developed technetium complexes, it includes relatively few rhenium complexes.

5.1 Bone-Targeting Diphosphonate Complexes

The most advanced along the path to clinical application are the diphosphonate complexes designed as palliative agents for relief of pain due to bone metastases in patients with cancer. This pain is believed to arise partly from an increase in intraosseous pressure due to the invading tumor, and partly because osteoclasts, the cells that induce dissolution of bone mineral and whose activity can become relatively uncontrolled in metastases, release various biochemical, mediators, and cytokines. Infiltrating inflammatory cells (lymphocytes) may also contribute to the release of these mediators. Both lymphocytes and osteoclasts are radiosensitive and respond to radiation doses not considered to be cytotoxic to tumor cells, and this renders pain relief by irradiation of the affected tissue feasible. It must be said, however, that neither the cause of pain nor the mechanisms of its relief by irradiation are well understood [123].

Radionuclides that become incorporated selectively into bone mineral in skeletal metastases can be exploited to provide this pain relief. ^{32}P-labeled polyphosphates and diphosphonates, which adsorb on the surface of bone mineral, have been used, and ^{89}Sr, a calcium analogue, is commercially available for treatment of bone pain. The possibility that ^{186}Re or ^{188}Re could be selectively delivered to bone was indicated by the development of ^{99m}Tc-labeled diphosphonate imaging agents [17,124]. These accumulate in bone by adsorption to the surface of bone mineral (hydroxyapatite and especially its precursor, amorphous low-density hydrated calcium phosphate), possibly with the diphosphonate ligand bridging between calcium ions in the crystal and technetium in a manner resembling **43** [125]. Although stability studies [126,127] have shown ^{186}Re–diphosphonates to bind to plasma protein increasingly with time, probably due to the *in vivo* decomposition of the complex, it is likely that the intact diphosphonate complex binds to the mineral component of bone. Increased blood flow and a higher fraction of amorphous calcium

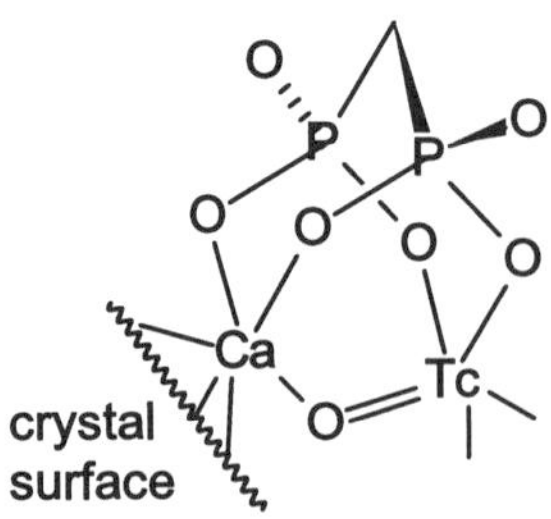

43

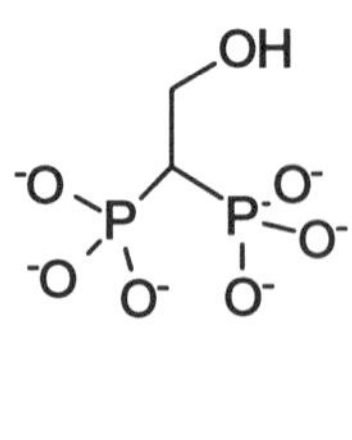

44

phosphate in diseased bone probably account for the lesion-to-normal bone selectivity. The mechanisms have been much more thoroughly investigated for the technetium–diphosphonate complexes [128], and the mechanisms for the rhenium complexes are expected to be similar.

Most work with rhenium–diphosphonate complexes has employed hydroxyethylidenediphosphonate (HEDP, **44**). Mathieu and co-workers [124] labeled HEDP by electrolysis, whereby two tin electrodes were inserted into a vial containing a solution of 3 mg HEDP and $^{186}ReO_4^-$ at pH 6, for 30 s, giving a 95% labeling yield after filtration. The unpurified mixture showed comparable skeletal and kidney excretion to the Tc analogue in rats at 4 h and 48 h post-injection. Stannous chloride was used as the reductant in the synthesis of Weininger and co-workers [129] giving improved stability to reoxidation. Deutsch and co-workers [125] describe use of greater concentrations of kit components (75 mg HEDP; 25 mg $SnCl_2{\cdot}2H_2O$; 116 mg sodium acetate, pH 6–7) in 1 mL $^{186}ReO_4^-$ (ca. 1 mg Re) and a longer reaction time (8 h at room temperature, anaerobic conditions). Concentration of rhenium in the preparation was found to affect yield of bone-targeting tracer [129]. Preparations with ^{188}Re at the no-carrier added level give complexes with very poor skeletal selectivity (Knapp F.F., Jr, personal communication), while a mixture of ^{188}Re (carrier-free) and ^{186}Re (not carrier-free) was effective [130]. These observations suggest that a variable molecular weight polymer/oligomer may be formed with bone affinity dependent on molecular weight. Consistent with this notion, anion exchange HPLC of the product showed a mixture of unidentified components whose respective yields varied with time and temperature, and whose bone selectivities in animals were not the same. The problem of high kidney uptake observed in rodents was, fortunately, not found in humans [131]. Further optimization led to a preparation in which purification could be effected without recourse to HPLC [125,132,133]. Ascorbic acid was included as an antioxidant, pH was optimized at 3–3.5, and high temperature was applied (100 °C for 10 min) to give >90% yield of HEDP complexes. The simplified purification involved loading the mixture onto an anion exchange column (prewashed with ascorbic acid), eluting with a solution of ascorbic acid to remove weakly bound components, and finally eluting the desired product (strongly bound, high bone affinity) with a saline/HEDP/ascorbic acid mixture at pH 5 to recover approximately 50% of the initial activity. Omission of a purification step altogether has been made possible by the development of a kit preparation procedure along these lines [134]. Further modifications include use of gentisic acid as antioxidant, and raising the reaction temperature still further (autoclaving at 120 °C). The latter modification appears to enhance bone:soft tissue ratios significantly [135]. A freeze-dried radioactive mixture was developed to minimize the problem of autoradiolysis, which is less prevalent in the absence of water. The radiopharmaceutical was prepared, purified and lyophilized to give a product that could be shipped and reconstituted with saline or water at the point of use. A disadvantage of this approach is the time taken for the lyophilization step, which results in considerable waste of radioactivity [136].

There is no firm identification of the structure of the active rhenium HEDP complexes in these preparations. However, EXAFS studies [137] of noncrystalline solid samples and frozen aqueous solutions of ReHEDP support the idea of oligomeric or polymeric complexes and suggest the presence of Re≡Re triple bonds and hence a rhenium oxidation state below (V). The Re is coordinated to 6 oxygen atoms (4 at 2 Å and 2 at 2.1 Å) probably from HEDP and water ligands, with the two longer distances possibly representing bridging oxygen atoms.

Clinical studies with ^{186}Re, and more recently with ^{188}Re [137a,b] have confirmed increased uptake of the radiopharmaceutical in bone metastases in humans, and have demonstrated a beneficial effect on pain. Typical results indicate that at least 20% of patients become pain-free and at least 70% experience some pain relief. Further, the onset of new bone, pain is delayed compared to external beam radiotherapy of localized metastases. These results are similar to those obtained with strontium-89 and other bone therapy agents under development [133,138–143]. The tumor to marrow-absorbed radiation dose ratios (22:1) are double those achieved using ^{89}Sr[139]. Nevertheless, bone marrow irradiation is significant and temporary drop in platelet and leucocyte numbers is a limiting side effect [144]. A transient "pain-flare" reaction (increase in pain) is seen in 50% of patients, occurring within 72 h postinjection, which subsides within 24–48 h.

5.2 Dimercaptosuccinic Acid Complex

A second rhenium-essential tracer, which also has its origins in diagnostic applications of analogous technetium complexes, is the rhenium(V) oxo-complex of *meso*-2,3-dimercaptosuccinic acid (DMSA) (Figure 3). This complex was designed as a β-emitting analogue of the radiopharmaceutical known as "pentavalent technetium-99m–DMSA" (^{99m}Tc(V)DMSA, distinguishing it from the routinely used kidney imag-

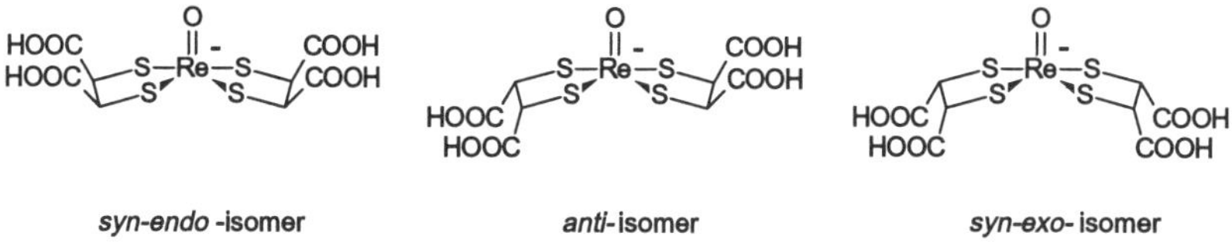

Figure 3. Isomers of rhenium dimercaptosuccinic acid complexes.

ing agent ^{99m}Tc–DMSA), which has found widespread use as an imaging agent for medullary thyroid carcinoma, a rare malignancy arising from C-cells in the thyroid. ^{99m}Tc(V)DMSA was recently identified as a mixture of the three isomers of the square pyramidal technetium(V) complex analogous to the complexes shown in Figure 3 [145]. As expected from the known structures of rhenium–oxo-bis-1,2-dithiolate complexes [10,52] the rhenium was found to be coordinated by four thiolate donors and an apical oxo-group. The ^{188}Re/^{186}Re-labeled complex was readily synthesized from the "kit" vials used to prepare the renal agent ^{99m}Tc–DMSA. This preparation illustrates the different redox behavior of rhenium and technetium. The kit has to be modified for production of Tc(V)DMSA by raising the pH from 3 to 8, and removing some of the stannous chloride [146] to prevent reduction below Tc(V). In contrast, synthesis of the rhenium complex requires no modification because rhenium is harder to reduce: acid conditions and excess stannous chloride are *required.* Simply adding ^{186}Re or ^{188}Re perrhenate and heating to 100 °C for 30 min provides the radiochemically pure rhenium complex [147,148]. It can also be synthesized from ^{188}Re-labeled $ReOCl_3(PPh_3)_2$ (obtained by PPh_3/HCl reduction of perrhenate in a two-phase solvent system) [149]. The isomeric composition of the complex was elucidated by a combination of HPLC, ^{1}H NMR, and X-ray crystallography [150], showing that the *syn-endo* and *anti* isomers are most abundant while the *syn-exo* complex contributes relatively little to the mixture. The composition can, however, be varied by altering the preparation conditions. The isomers are interconverted by acid-catalyzed ligand dissociation slowly (over a period of weeks) at physiological pH [150].

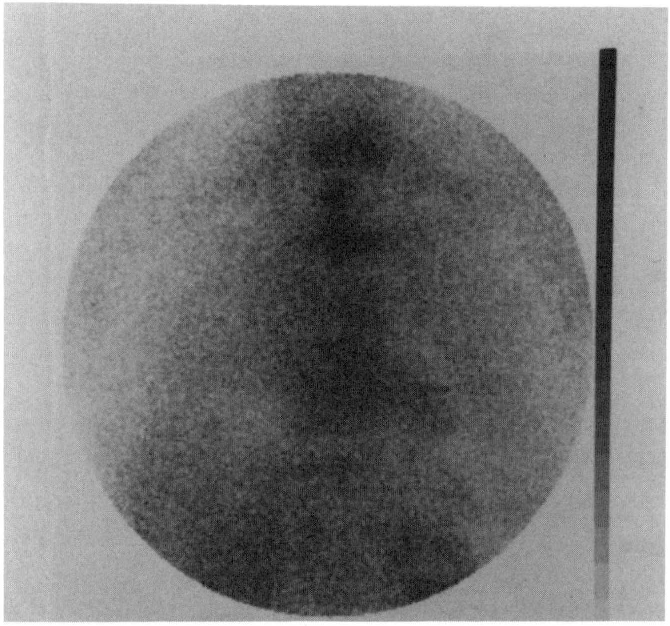

Figure 4. Gamma camera image of a patient with medullary thyroid carcinoma showing selective uptake of Re-186-DMSA in a tumor at the base of the neck (taken 24 h after injection).

The biological properties of ^{188}Re(V)DMSA have been investigated in humans and in animals. In patients with medullary thyroid carcinoma, it shows selective uptake in tumor tissue (Figure 4) similar to that of the technetium analogue [151], offering a possibility for targeted radiotherapy of this disease. It is also taken up selectively in bone metastases in cancer patients (Figure 5), probably by adsorption to the calcium-rich surface of bone mineral via the carboxylate and oxo groups [152]. This offers the potential for palliative therapy. HPLC studies of blood and urine from patients injected with the complex show no evidence for decomposition to perrhenate or any other chemical form of rhenium over 24 h, suggesting that the targeting properties are those of the intact complex and not a metabolite or breakdown product [152]. It has been claimed that use of dithionite in place of $SnCl_2$ in the preparation of ^{188}Re(V)DMSA gives reduced renal uptake while maintaining bone affinity [152a].

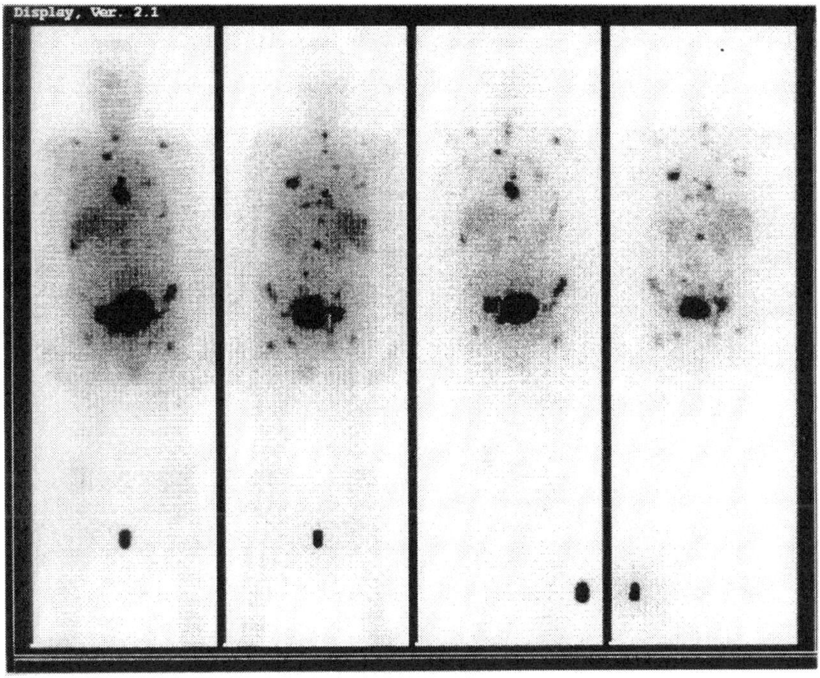

Figure 5. Gamma camera image of a prostate cancer patient with widespread bone metastases, showing uptake of Re-188-DMSA in bone metastases. Left to right: anterior, 3 h; posterior, 3 h; anterior, 24 h; posterior, 24 h.

5.3 Steroid Analogues

An ambitious approach to rhenium- (and technetium-) essential tracers is being developed by Katzenellenbogen and co-workers. The aim is to use the square-pyramidal ReO^{3+} core as the framework for assembly of complexes resembling steroids. Preliminary work has shown that the bis-chelate ring structure of the complex is a structural mimic of the decalin-like BC ring system of the steroid (Figure 6). Using various 2-aminothiols to chelate the rhenium, stable complexes containing two different ligands were synthesized from mixtures of the ligands; indeed, the heteroleptic complexes, such as **45** and **46** (Figure 6) with sulfur donors mutually *trans*, appeared to be preferred. By careful design of the ligands to include five- and six-membered rings, it is hoped that synthesis of specific hormone mimics can be achieved [153,154].

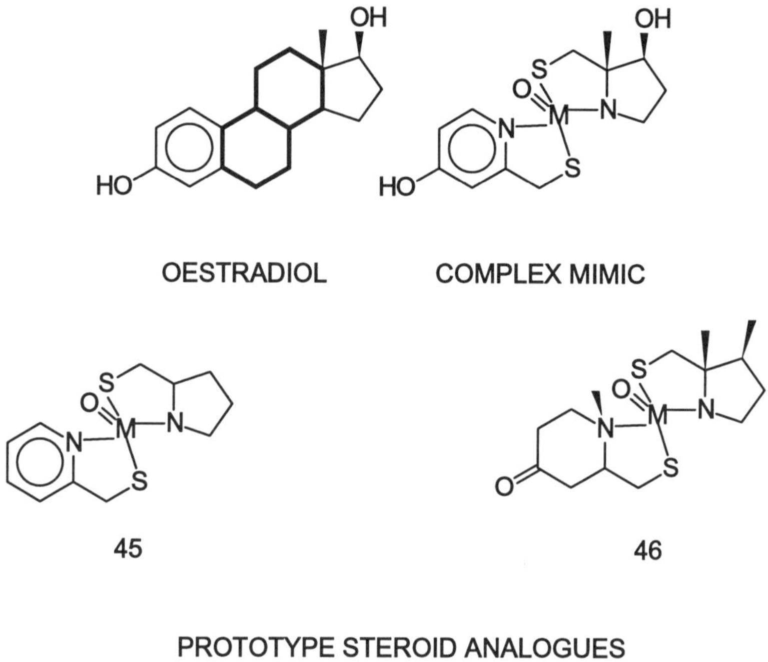

Figure 6. Construction of steroid structural analogues based on the square pyramidal ReO^{3+} core.

6. PARTICULATE DELIVERY AGENTS

Particulate delivery agents of various sizes (<1 μm to >40 μm in diameter) labeled with radiotracers are useful in several ways in nuclear medicine: as targeting agents they can be given systemically to target lung (by trapping of large particles in the capillary network) or the reticuloendothelial system (liver Kupffer cells, bone marrow) by phagocytosis. They can also be used as a relatively immobile form of radioactivity that can be directly injected into a lesion and expected to remain there. This is exploited in treatment of inflamed synovia in arthritic conditions by direct injection into the joint (radiation synovectomy), and for treatment of various tumors by direct injection into the tumor or the body cavity containing the tumor (endocavitary irradiation). Finally, the ability of large particles to become trapped in the capillary bed (radioembolization) can be exploited for tumor treatment where injection into an artery supplying the tumor is feasible. The most important use of the latter modality is in treatment of hepatocellular carcinoma by injection into the hepatic artery.

Various methods for incorporating radioactive rhenium into suitable particles have been introduced. The earliest example is the use of ^{186}Re-labeled sulfur colloid for radiation synovectomy. Commercial preparations for this application have been available for many years (CIS, France, 1987 catalog). ^{186}Re–sulfur colloid has been used for endocavitary irradiation of cystic craniopharyngiomas, where irradiation of surrounding tissue is a risk that must be avoided [155].

Hydroxyapatite particles can be used for radiation synovectomy by exploiting the strong binding of rhenium diphosphonate complexes to the surface of hydroxyapatite (*vide supra*). The same ^{186}Re-HEDP preparations as for bone palliative therapy are used. When injected into joints, these particles (mean diameter 25 μm, maximum diameter 45 μm) remain within the joint to the extent of at least 95% for several days in arthritic rabbits and rats [156]. More recently, microspheres have been labeled with ^{188}Re for this purpose [156a,b].

^{186}Re has been incorporated into biodegradable microcapsules formed by polymerization of isobutylcyanoacrylate in the presence of ^{186}Re dispersed in organic solvent. With a mean diameter of 10–15 μm these particles were used for radioembolization therapy of B16 melanoma induced in mice. More than 90% of the injected radioactivity was trapped within the tumors and tumor growth retardation was observed [157].

Particle suspensions labeled with rhenium radioisotopes can be produced by preparation of liposomes in the presence of the radioisotope. Liposomes (70 nm diameter) are targeted to the reticuloendothelial system, and hence could be used for treatment of hepatocellular carcinoma by exploiting trapping in the liver Kupffer cells. Two routes for the preparation are feasible: highly lipophilic complexes can be attached by incorporation into the lipid bilayer membrane, or hydrophilic complexes can be trapped in the aqueous phase inside the liposome [158]. The former has been achieved by using ^{186}Re-labeled $ReOCl_3(PPh_3)_2$ as the lipophilic species, which is added to the lipid preparation prior to liposome formation by detergent removal [158].

Iodine-131-labeled Lipiodol, a polyiodinated poppy seed oil which becomes "particulate" on dispersion in aqueous media, has been used for some time in treatment of hepatocellular carcinoma by arterial injection. This targeting approach has been coupled with the superior radioactivity properties of rhenium radioisotopes by exploiting the lipid solubility of the diamino–dithiolate (**16**) complexes to incorporate them into the lipiodol [62].

7. PRACTICAL RADIOCHEMISTRY CONSIDERATIONS

Several limitations on the synthetic techniques that can be employed are imposed by the need for rapidity and minimization of handling because of the radiation hazard, and the low concentration and small physical quantities of the compounds. Purification steps should be eliminated if possible by optimizing yields. Where purification is unavoidable, simple procedures are employed such as use of anion exchange columns to remove perrhenate (the most common contaminant in the final product). A variety of disposable sample preparation columns are well suited to this purpose and are available containing small quantities of anion or cation exchange materials (0.1 to 0.5 g typically) such as quaternary ammonium-, primary ammonium-, or sulfonate-derivatized silica. Reversed phase columns are also often used (C8 or C18-derivatized silica). The purification is often thus reduced to a simple "filtration" step which can be performed aseptically.

The need to achieve high yield in "one-pot" synthesis, coupled to the relative kinetic inertness of rhenium complex (e.g. compared to technetium) and the mild conditions required has led to the development of useful versatile rhenium(V) intermediates that can be quickly prepared in quantitative yield, and are metastable, i.e. kinetically labile enough to react rapidly with the final chelator, again in high yield. The most widely used ligands suitable for this purpose are polydentate hydroxycarboxylic acids such as glucoheptonate [116a], citrate (**47**), tartrate (**48**), and 2-hydroxyisobutyric acid (**49**) [159]. Examples are discussed elsewhere in this chapter. They are typically used in the presence of Sn(II) to reduce Re(VII) to Re(V), at moderately elevated temperature (50–100 °C) at pH 2–3 (acid pH promotes reduction of perrhenate, presumably by facilitat-

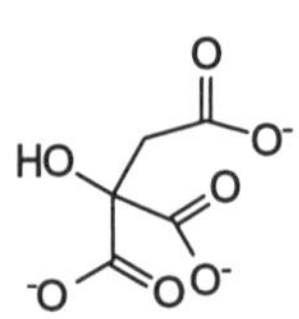

47

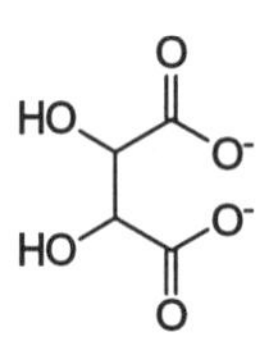

48

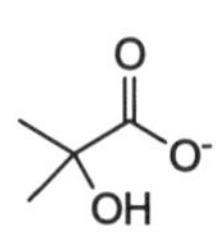

49

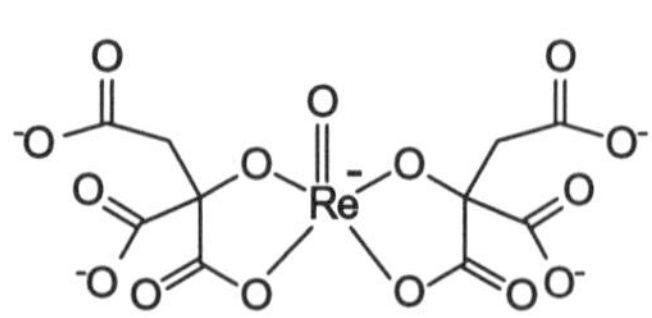

50

ing removal of π-donor oxo ligands as water). The subsequent transchelation step can then be carried out at or below room temperature at less acidic, or alkaline pH. The structure of the intermediate complexes is unknown but is presumed to contain rhenium in oxidation state (V), possibly as a bis-complex of the type represented by **50**: 2-hydroxycarboxylates are known to chelate the rhenium oxo center.

Other approaches to reduction of perrhenate to a suitable Re(V) intermediate include the use of triphenylphosphine and concentrated hydrochloric acid in a mixed solvent system [149]. This has led to an effective synthesis of the rhenium dimercaptosuccinic acid complex, and could prove useful in synthesis of other Re(V) complexes. It is presumed that the intermediate organic-soluble complex is the $ReOCl_3(PPh_3)_2$ species well known to synthetic inorganic chemists [149]. Other agents investigated for reduction of radioactive perrhenate include dithionite and hypophosphorous acid [50], the latter without success. Electrochemical reduction of perrhenate in hydrochloric acid has been investigated as a method of producing Re(V) (presumably $ReOCl_4^-$) intermediates. Good yields of ^{188}Re(V) were obtained but labeling of protein fragments was not effective, possibly because continuous exposure of reduced rhenium to reducing agent is required to prevent reoxidation to Re(VII) before incorporation of the rhenium into the final stable chelate [160]. Use of water-soluble phosphines such as sulfonated triphenylphosphines has been investigated as a reducing agent for perrhenate in water, without success (S. Prakash and P.J. Blower, unpublished). This is probably because phosphines are only effective reducing agents in nonaqueous solvents, not for solubility reasons but because reduction of perrhenate is favored by the absence of water. It is thus often useful to switch to a nonaqueous solvent, and to do this simple minimal-handling methods, such as extraction into methyl ethyl ketone [17], are preferable to evaporation. Alternatively, perrhenate can be extracted onto an anion exchange material using the column system shown in Figure 1 [34]. The perrhenate can be removed from the dried resin with nonaqueous solvent, or the chelation reaction can be carried out on the column in nonaqueous solvent, and the product subsequently eluted with an aqueous medium suitable for injection.

An additional problem faced by those synthesizing therapeutic doses (i.e. very high activities, of the order of tens of gigabequerels) is autoradiolysis, or decomposition of the product complex as a result of the action of radical species formed by the interaction of the emitted electrons with water. This is compounded by the relative ease of oxidation of reduced rhenium to perrhenate under aerobic conditions. The expedient solution to this problem is addition of nontoxic radical scavengers, most commonly ascorbic acid and gentisic acid [152,161,162]. In addition, so called "challenging agents" are added, with the intention of complexing redox active contaminant metal ions which may engage in redox reactions with the product (e.g. DTPA or EDTA to complex Fe^{2+}/Fe^{3+}) [161]. These additives have been successful in stabilizing both radiolabeled antibody preparations and small molecular complexes of ^{186}Re and ^{188}Re such as Re-HEDP (*vide supra*) and Re-DMSA [152], and also in improving the elution yield of ^{188}Re generators by including ascorbic acid in the eluent at concentrations as low as 0.01% [44]. Human serum albumin, added to preparations at the end of the synthesis and purification, has been shown to enhance the stabilizing effects of the aforementioned stabilizing agents [161].

CONCLUSION

Despite the large body of work described above, and the recent commercial introduction of ^{186}Re-labeled diphosphonates for palliative radiotherapy, it has to be said that the prospect of *routine* use of rhenium radioisotopes for targeted *curative* treatment of cancer is still some way off. This is true of radioisotopes of other elements too. To date, the only targeted radionuclide treatment for cancer is use of ^{131}I-iodide for thyroid cancer. This has been available for several decades and stands as the example that proves the principle of targeted radionuclide therapy. That other examples may soon emerge is heralded by the recent dramatic success with ^{131}I-labeled *meta*-iodobenzylguanidine (mIBG), which is as effective as chemotherapy in presurgical reduction of tumor volume in children with neuroblastoma, but causes none of the debilitating side effects [163]. Routine application is delayed by the necessarily slow pace at which new treatments can be tested in patients—^{131}I-mIBG was first developed in the early 1980s. That ^{131}I is the only therapeutic radionuclide in routine use is not due to any advantages of the radionuclide: indeed, it has many serious disadvantages compared to the rhenium radioisotopes. Rather, it is because the ^{131}I agents were initially developed many years ago when the physically superior metallic radionuclides were not readily available. Although many of the initiatives described in this chapter are at an early stage of development, given time for clinical evaluation in patients *in the early stages of disease* it seems inevitable that delivery systems as selective as mIBG will be developed to exploit the new radionuclides, leading to more effective and less debilitating cancer treatments.

ACKNOWLEDGMENT

The authors thank Dr. F.F. (Russ) Knapp, Jr. for communicating results to us prior to publication.

REFERENCES

[1] Wessels, B.W. and Rogus, R.D., Med. Phys., 11 (1984) 638–645.
[2] O'Donoghue, J.A., Bardies, M. and Wheldon, T.E., J. Nucl. Med., 36 (1995) 1902–1909.
[3] Simpkin, D.J. and Mackie, T.R., Med. Phys., 17 (1990) 179.
[4] Yorke, E.D., Beaumier, P.L., Wessels, B.W., Fritzberg, A.R. and Morgan, A.C. Jr., Nucl. Med. Biol., 18 (1991) 827.
[5] Liu, Y.-F., Tang, R.-H., Zhang, Q.-X., Shi, J.-Y., Li, X.-M., Liu, Z.-Q. and Zhao, W., Biol. Trace Element Res., 9 (1986) 89–99.
[6] Deutsch, E., Libson, K. and Vanderheyden, J.-L., in Nicolini, M., Bandoli, G. and Mazzi, U. (eds.), *Technetium and Rhenium in Chemistry and Nuclear Medicine 3*, Raven Press, New York, 1990, pp. 13–22
[6a] Kotzerke, J., Fenchel, S., Guhlmann, A., Stabin, M., Rentschler, M., Knapp, F.F. Jr. and Reske, S.N., Nucl. Med. Commun., 19 (1998) 795–801.
[7] Taurog, A., in Braverman, L.E. and Utiger, R.D. (eds.), *The Thyroid*, 6th ed, Lippincott, Philadelphia, 1991.
[8] Registry of Toxic Effects of Chemical Substances, 2 (1979).
[9] Nielson, K.B., Atkin, C.L. and Winge, D.R., J. Biol. Chem. 260 (1985) 5342–5348.
[10] Blower, P.J., Dilworth, J.R., Hutchinson, J.P., Nicholson, T. and Zubieta, J., J. Chem. Soc., Dalton Trans., (1996) 1339–1345.
[11] Chatt, J. and Rowe, G.A., J. Chem. Soc.,(1962) 4019–1033.
[12] Libson, K., Helm, L., Roodt, A., Cutler, C., Merbach, A.E., Sullivan, J.C. and Deutsch, E., in Nicolini, M., Bandoli, G. and Mazzi, U. (eds.), *Technetium and Rhenium in Chemistry and Nuclear Medicine 3*, Raven Press, New York, 1990, pp. 31–33.
[13] Helm, L., Deutsch, K., Deutsch, E.A. and Merbach, A.E., Helv. Chim. Acta, 75 (1992) 210–217.
[14] Edwards, D.S., Cheesman, E.H., Watson, M.W., Maheu, L.J., Nguyen, S.A., Dimitre, L., Nason, T., Watson, A.D. and Walovitch, R., in Nicolini, M., Bandoli, G. and Mazzi, U. (eds.), *Technetium and Rhenium in Chemistry and Nuclear Medicine 3,* Raven Press, New York, 1990, pp. 434–444.
[15] Dilworth, J.R., Archer, C.M., Bishop, P.T., Kelly, J.D. and Higley, B., J. Nucl. Med. 30 (1989) 773 (abstract).
[16] Majali, M.A., J. Radioanal. Nucl. Chem. Articles, 170 (1993) 471–477.
[17] Eisenhut, M., Int. J. Appl. Radiat. Isot., 33 (1982) 99–103.
[18] Vanderheyden, J.-L., Su, F.-M. and Ehrhardt, G.J., US Patent 5,053,186 (1991).

[19] Vanderheyden, J.-L., Hegg, M.J. and Deutsch, E., Inorg. Chem., 24 (1985) 1666–1673.
[20] Hafeli, U., Tiefenauer, L., Schubiger, P.A. and Weder, H.G., Nucl. Med. Biol. ,18 (1991) 449.
[21] Hafeli, U., Tiefenauer, L., Schubiger, P.A. and Weder, H.G., in Nicolini, M., Bandoli, G. and Mazzi, U. (eds.), *Technetium and Rhenium in Chemistry and Nuclear Medicine 3*, Raven Press, New York, 1990, pp. 643–646.
[22] Schwarzbach, R., Blauenstein, P., Jege, J. and Schubiger, P.A., J. Labelled Compounds Radiopharm., 37 (1995) 816 (abstract).
[23] Griffiths, G.L., Goldenberg, D.M., Sharkey, R.M., Knapp, F.F. Jr., Callahan, A.P., Tejada, G. and Hansen, H.J., Radioact. Radiochem., 3 (1992) 33–37.
[24] Knapp, F.F. Jr., Eur. J. Nucl. Med., 21 (1994) 1151–1165.
[25] Kodina, G., Tulskaya, T., Gureev, E., Brodskaya, G., Gapurova, O. and Drosdovsky, B., in Nicolini, M., Bandoli, G. and Mazzi, U. (eds.), *Technetium and Rhenium in Chemistry and Nuclear Medicine 3*, Raven Press, New York, 1990, pp. 635–641.
[26] Mikheev, N.B., Popovich, V.B., Rumer, I.A., Savel'ev, G.I. and Volkova, N.L., Isotopenpraxis, 8 (1972) 248–251.
[27] Botros, N., El-Garhy, M., Abdulla, S. and Aly, H.F., Isotopenpraxis, 22 (1986) 368–371.
[28] Callahan, A.P., Rice, D.E., Knapp, F.F. Jr., Nuc. Compact, 20 (1989) 3–6.
[29] Brodskaya, G.A. and Gapurova, O.U., Radiokhimiya, 35 (1993) 92–97.
[30] Knapp, F.F. Jr., Callahan, A.P., Beets, A.L., Mirzadeh, S. and Hsieh, B.-T., Appl. Radiat. Isot., 45 (1994) 1123–1128.
[31] Callahan, A.P., Rice, D.E., McPherson, D.W., Mirzadeh, S., Knapp, F.F. Jr., Appl. Radiat. Isot. 43 (1992) 801–804.
[32] Vanderheyden, J.-L. and Su, F.-M., US Patent 4,990,787 (1991).
[33] Singh, J., Reghebi, K., Lazarus, C.R., Clarke, S.E.M., Callahan, A.P., Knapp, F.F. Jr. and Blower, P.J., Nucl. Med. Commun., 14 (1993) 197.
[34] Blower, P.J., Nucl. Med. Commun., 14 (1993) 995–997.
[35] Kamioki, H., Mirzadeh, S., Lambrecht, R.M., Knapp, R. Jr. and Dadachova, K., Radiochimica Acta, 65 (1994) 39–46.
[36] Lisic, E.C., Callahan, A.P., Mirzadeh, S. and Knapp, F.F. Jr., Radioact Radiochem., 3 (1992) 42–45.
[37] Knapp, F.F. Jr., Lisic, E.C., Mirzadeh, S. and Callahan, A.P., US Patent 5,186,913 (1993).
[38] Knapp, F.F. Jr., Lisic, E.C., Mirzadeh, S. and Callahan, A.P., International Patent WO 92/19541 (1992).
[39] Malyshev, K.V. and Smirnov, V.V., Radiokhimiya, 17 (1975) 249–251.
[40] Ehrhardt, G.J., Vanderheyden, J.-L. and Su, F.-M., International Patent WO 88/03697 (1988).
[41] Ehrhardt, G.J., Ketring, A.R., Turoin, T.A., Razavi, M.-S., Vanderheyden, J.-L., Su, F.-M. and Fritzberg, A.R., in Nicolini, M., Bandoli, G., Mazzi, U. (eds.), *Technetium and Rhenium in Chemistry and Nuclear Medicine 3*, Raven Press, New York, 1990, pp. 631–634.
[42] Ehrhardt, G.J., Ketring, A.R. and Liang, Q., Radioact. Radiochem., 3 (1992) 38–41.

[43] Ehrhardt, G.J., Wolfangel, R.G. and Deutsch, E., International Patent WO 94/04463 (1994).
[44] Hsieh, B.-T., Callahan, A.P., Beets, A.L., Ting, G. and Knapp, F.F. Jr., Appl. Radiat. Isot., 47 (1996) 23–26.
[44a] Knapp, F.F., Mirzadeh, S., Beets, A.L., O'Doherty, M.J., Blower, P.J., Verdera, E.S., Gaudiano, J.S., Kropp, J., Guhlke, S., Palmedo, H. and Biersack, H.J., Appl. Radiat. Isot., 49 (1998) 309–315.
[45] Fritzberg, A.R., Berninger, R.W., Hadley, S.W. and Wester, D.W., Pharmaceutical Res., 5 (1988) 325–334.
[46] Mears, C.F., in Feeney, R.E. and Whitaker, J.R. (eds.), *Protein Tailoring for Food and Medical Uses*, Marcel Dekker, New York, 1986.
[47] Hermanson, G.T., *Bioconjugate Techniques*, Academic Press, San Diego, USA, 1996.
[48] Hnatowich, D.J., Nucl. Med. Biol., 13 (1986) 353–358.
[49] Childs, R.L. and Hnatowich, D.J., J. Nucl. Med., 26 (1985) 293–299.
[50] Quadri, S.M. and Wessels, B.W., Nucl. Med. Biol., 13 (1986) 447.
[51] Koros, A.M.C., Tobin, M.J., Epperly, Levine, G. and McGinley, J.R., Anticancer Res., 13 (1993) 1953–1956.
[52] Davison, A., Orvig, C., Trop, H.S., Sohn, M., DePamphilis, B.V. and Jones, A.G., Inorg. Chem., 19 (1980) 1988–1992.
[53] Davison, A., Jones, A.G., Orvig, C. and Sohn, M., Inorg. Chem., 20 (1981) 1629–1632.
[54] Brenner, D., Davison, A., Lister-James, J. and Jones, A.G., Inorg. Chem. 23 (1984) 3793–3797.
[55] Kasina, S., Fritzberg, A.R., Johnson, D.L. and Eshima, D., J. Med. Chem., 29 (1986) 1933–1940.
[56] Fritzberg, A.R., Kasina, S., Srinivasan, A. and Wilbur, D.S., US Patent 4,897,255 (1990).
[57] Vanderheyden, J.-L., Rao, T.N., Kasina, S., Wester, D., Su, F.-M. and Fritzberg, A.R., in Nicolini, M., Bandoli, G. and Mazzi, U. (eds.), *Technetium and Rhenium in Chemistry and Nuclear Medicine 3*, Raven Press, New York, 1990, pp. 623–629.
[58] Rao, T.N., Brixner, D.I., Kasina, S., Srinivasan, D.W., Vanderheyden, J.-L. and Fritzberg, A.R., Abstracts of 1990 ACS meeting, 1990 abstract no. 185 INOR.
[59] Rao, T.N., Brixner, D.I., Srinivasan, A., Kasina, S., Vanderheyden, J.-L., Wester, D.W. and Fritzberg, A.R., Appl. Radiat. Isot., 2 (1991) 525–530.
[60] Fritzberg, A.R., Abrams, P.G., Beaumier, P.L., Kasina, S., Morgan, A.C. Jr., Rao, T.N., Reno, J.R., Sanderson, J.A., Srinivasan, A., Wilbur, D.S. and Vanderheyden, J.-L., Proc. Natl. Acad. Sci. 85 (1988) 4025–4029.
[61] Mahmood, A., Baidoo, S.Z., Lever, S.Z., in Nicolini, M., Bandoli, G. and Mazzi, U. (eds.), *Technetium and Rhenium in Chemistry and Nuclear Medicine 3*, Raven Press, New York, 1990, pp. 119–124.
[62] Jackson, T.W., Kojima, M. and Lambrecht, R.M., European Patent 582,892 (1994).
[63] DiZio, J.P., Anderson, C.J., Davison, A., Ehrhardt, G.J., Carlson, K.E., Welch, M.J. and Katzenellenbogen, J.A., J. Nucl. Med., 33 (1992) 558.

[64] DiZio, J.P., Fiaschi, R., Davison, A., Jones, A.G. and Katzenellenbogen, J.A., Bioconjugate Chem., 2 (1992) 353–366.

[65] O'Neil, J.P., Carlson, K.E., Anderson, C.J., Welch, M.J. and Katzenellenbogen, J.A., Bioconjugate Chem., 5 (1994) 182–193.

[66] Fritzberg, A.R., Kasina, S., Eshima, D. and Johnson, D.L., J. Nucl. Med., 27 (1986) 111–116.

[67] Nosco, D.L. and Manning, R.G., J. Nucl. Med., 27 (1986) 939 (abstract).

[68] Inorg. Chim. Acta 180 (1991) 63–67.

[69] Wilkening, D., Srinivasan, A., Kasina, S., Rao, T.N., Vanderheyden, J.-L. and Fritzberg, A.R., J. Radiolabelled Compounds Radiopharm., 25 (1988) 261–263.

[70] Goldrosen, M.H., Biddle, W.C., Pancook, J., Bakshi, S., Vanderheyden, J.-L., Fritzberg, A.R., Morgan, A.C. Jr. and Foon, K.A., Cancer Res., 50 (1990) 7973–7978.

[71] Beaumier, P.L., Venkatesan, P., Vanderheyden, J.-L., Burgua, W.D., Kunz, L.L., Fritzberg, A.R., Abrams, P.G. and Morgan, A.C. Jr., Cancer Res., 51 (1991) 676–681.

[72] Langmuir, V.K., Mendonca, H.L., Vanderheyden, J.-L. and Su, F.-M., Int. J. Radiat. Oncol. Biol. Phys. 24 (1992) 127–132.

[73] Beaumier, P.L., Vanderheyden, J.-L., Venkatesan, P., Kunz, L., Duke, S.P., Abrams, P.G., Morgan, A.C. Jr. and Fritzberg, A.R., Antibody Immunoconjugates Radiopharm, 4 (1991) 735–744.

[74] Breitz, H.B., Weiden, P.L., Vanderheyden, J.-L., Appelbaum, J., Bjorn, M.J., Fer, M.F., Wolf, S.B., Ratliff, B.A., Seiler, C.A., Fossie, D.C., Fisher, D.R., Schroff, R.W., Fritzberg, A.R. and Abrams, P.G., J. Nucl. Med., 33 (1992) 1099–1109.

[75] Breitz, H.B., Fisher, D.R., Weiden, P.L., Durham, J.S., Ratliff, B.A., Bjorn, M.J., Beaumier, P.L. and Abrams, P.G., J. Nucl. Med., 34 (1993) 908–917.

[76] Weiden, P.L., Wolf, S.B., Breitz, H.B., Appelbaum, J.W., Seiler, C.A., Mallet, R., Bjorn, M.J., Su, F.-M., Fer, M.F. and Salk, D., Cancer (Suppl.), 73 (1994) 1093–1097.

[77] Schroff, R.W., Weiden, P.L., Appelbaum, J., Fer, M.F., Breitz, H., Vanderheyden, J.-L., Ratliff, B.A., Fisher, D., Foisie, D., Hanelin, L.G., Morgan, A.C. Jr., Fritzberg, A.R. and Abrams, P.G., Antibody, Immunoconjugate and Radiopharmaceuticals 3, (1990) 99–111.

[78] Jacobs, A.J., Fer, M., Su, F.-M., Breitz, H.B., Thompson, J., Goodgold, H., Cain, J., Heaps, J. and Weiden, P., Obstetrics and Gynaecology, 82 (1993) 586–593.

[79] Breitz, H.B., Durham, J.S., Fisher, D.L., Weiden, P.L., DeNardo, G.L., Goodgol, H.M. and Nelp, W.B., J. Nucl. Med. 36 (1995) 754–761.

[80] Weiden, P.L., Breitz, H.B., Seiler, C.A., Bjorn, M.J., Ratliff, B.A., Mallett, R., Beaumier, P.L., Appelbaum, J.W., Fritzberg, A.R. and Salk, D., J. Nucl. Med., 34 (1993) 2111–2119.

[81] Gustavson, L., Axworthy, D., Sanderson, J., Su, F.M., Srinivasan, A., Beaumier, A., Fritzberg, A. and Reno, J., Abstracts of Papers of the American Chemical Society Meeting 1992, paper 23-MEDI.

[82] Fritzberg, A., Srinivasan, A., Vanderheyden, J.-L., Kasina, S., Su, F.-M., Axworthy, D., Fitzner, J., Reno, J. and Beaumier, P., Antibody Immunoconjugates Radiopharm., 4 (1991) 216 (abstract).

[83] Gerretsen, M., Visser, G.W.M., Van Walsum, M., Meijer, C.J.L.M., Snow, G.B. and Van Dongen, G.A.M.S., Cancer Res., 53 (1993) 3524–3529.
[84] Visser, G.W.M., Gerretsen, M., Herscheid, J.D.M., Snow, G.B. and Van Dongen, G., J. Nucl. Med., 34 (1993) 1953–1963.
[85] Ram, S. and Buchsbaum, D.J., Cancer (Suppl.), 73 (1994) 769–773.
[86] Guhlke, S., Diemann, D., Zamora, P.O., Knapp, F.F. and Biersack, H.J., in Nicolini, M., Bandoli, G. and Mazzi, U. (eds.), *Technetium and Rhenium in Chemistry and Nuclear Medicine 4*, Proceedings of the symposium held in Bressanone, Italy, Sep 1994. Ed., SGEditoriali, Italy, 1995, pp. 363–366.
[87] Schaffland, A., Guhlke, S., Knapp, F.F. Jr., Zamora, P.O. and Biersack, H.J., J. Labelled Comp. Radiopharm., 37 (1995) 493–495 (abstract).
[88] Srinivasan, A., World Patent WO 93/15771 (1993).
[89] Schmidt, M.A., Grummon, G.D., Bugaj, J.E., Cannon, E.P., Cooke, I., Rosik, L.O., Deutsch, K.F. and Dunn, T.J., J. Nucl. Med., 35 (1994) 139 (abstract).
[89a] Al-Jeboori, M.J., Dilworth, J.R. and Zheng, Y.F., J. Chem. Soc. Dalton. Trans., (1998), 3215–3218.
[89b] Glaser, M., Howard, M.J., Howland, K., Powell, A.K., Rae, M.T., Wocadlo, S., Williamson, R.A. and Blower, P.J., J. Chem. Soc. Dalton Trans., (1998) 3087–3092.
[89c] Al-Jeboori, M.J., Dilworth, J.R. and Hiller, W., Inorg. Chim. Acta, 285 (1999) 76–80.
[90] Parker, D. and Roy, P.S., Inorg. Chem., 27 (1988) 4127–4130.
[91] Luna, S.A., Bolzati, C., Duatti, A., Zucchini, G.L., Bandili, G. and Refosco, F., Inorg. Chem. ,31 (1992) 2595–2598.
[92] Prakash, S., Went, M.J. and Blower, P.J., Nucl. Med. Biol., 23 (1996) 543–549.
[93] Huber, G.J., Alberto, R.A., Blauenstein, P. and Anderegg, G., J. Chem. Soc., Chem. Commun., (1989) 879–881.
[94] Salmain, M., Gunn, M., Gorfti, A., Top, S. and Jaouen, G., Bioconjugate Chem., 4 (1993) 425–433.
[95] Top, S., Vessieres, A. and Jaouen, G., J. Chem. Soc., Chem. Commun. (1994) 453–454.
[96] Spradau, T.W. and Katzenellenbogen, J.A., Organometallics, 17 (1998) 2009–2017.
[96a] Minutolo, F. and Katzenellenbogen, J.A., J. Am. Chem. Soc., 120 (1998) 4514–4515.
[97] Alberto, R., Schibli, R., Egli, A., Schubiger, P.A., Herrmann, W.A., Artus, G., Abram, U. and Kaden, T.A., J. Organomet. Chem., 493 (1995) 119–127.
[98] Alberto, R., Schibli, R., Schubiger, P.A., Abram, U. and Kaden, T.A., Polyhedron, 15 (1996) 1079–1089.
[98a] Egli, A., Hegetschweiler, K., Alberto, R., Abram, U., Schibli, R., Hedinger, V., Gramlich, R., Kissner, P. and Schibiger, P.A., Organometallics, 16 (1997) 1833–1840.
[99] Kuppers, H.-J., Nuber, B., Weiss, J. and Cooper, S.R., J. Chem. Soc., Chem. Commun., (1990) 979–980.
[100] Matondo, S.O.C., Mountford, P., Watkin, D.J., Jones, W.B. and Cooper, S.R., J. Chem. Soc., Chem. Commun., (1995) 161–162.

[101] Mullen, G.E.D., Went, M.J., Wocadlo, S., Powell, A.K., Blower, P.J., Angew. Chem. Int. Ed. 36 (1997) 1205–1207.

[102] Najafi, A., Alauddin, M.M., Sosa, A., Ma, G.Q., Chen, D.C.P., Epstein, A.L. and Siegel, M.E., Nucl. Med. Biol. 19 (1992) 205.

[103] Papadopoulos, M., Pirmettis, I., Varvarigou, A., Terzis, A. and Chiotellis, E., J. Nucl. Med., 35 (1994) 244 (abstract).

[104] Pirmettis, I., Papadopoulos, M., Tsoukalas, C., Raptopolou, C.P., Terzis, A. and Chiotellis, E., J. Labelled Compounds Radiopharm., 37 (1995) 447–449 (abstract).

[105] Spies, H., Johannsen, B., Pietzsch, H.-J., Noll, B., Scheunemann, M., Fietz, T., Berger, R., Brust, P. and Leibnitz, P., J. Labelled Compounds Radiopharm. 37 (1995) 319–320 (abstract).

[105a] Maresca, K.P., Femia, F.J., Babich, J.W. and Zubieta, J., Inorg., Chem. Commun., 1 (1998) 209–212.

[105b] Papadopoulos, M., Pirmetti, I., Raptopolou, C., Chiotellis, E., Friebe, M., Berger, R., Spies, H. and Johannsen, B., Appl. Radiat. Isot., 49 (1998) 961–963.

[105c] Hoepping, A., Brust, P., Berger, R., Leibnitz, P., Spies, H., Machill, S., Scheller, D. and Johannsen, B., Bioorg. Med. Chem., 6 (1998) 1663–1672.

[105d] Wust, F., Scheller, D., Spies, H. and Johannsen, B., Eur. J. Inorg. Chem., (1998) 789–793.

[106] Wu, R.S., European Patent 0,306,168 (1989).

[107] Kasina, S., Srinivasan, A., Sanderson, J. and Fritzberg, A.R., US Patent 5,175,256 (1992).

[108] Francesconi, L.C. and Treher, E.N., European Patent 0 311 891 A1 (1989).

[109] Jurisson, S., Linder, K.E., Hirth, W., Feld, T., Nowotnik, D.P., Nunn, A.D. and Eckelman, W.C., in Nicolini, M., Bandoli, G. and Mazzi, U. (eds.), *Technetium and Rhenium in Chemistry and Nuclear Medicine 3*, Raven Press, New York, 1990, pp. 189–193.

[110] Kopcka, K., Katti, K.V., Ketring, A.R. and Volkert, W.A., J. Nucl. Med., 33 (1992) 979 (abstract).

[111] Katti, K.V., Volkert, W.A., Ketring, A.R. and Singh, P.R., World Patent WO 93/08839 (1993).

[112] Griffiths, G.L., European Patent 0 403 225 (1990).

[113] Hansen, H.J. and Griffiths, G.L., US Patent 5,328,679 (1994).

[114] Griffiths, G.L., Goldenberg, D.M., Knapp, F.F. Jr., Callahan, A.P., Chang, C.-H. and Hansen, H.J., Cancer Res., 51 (1991) 4594–4602.

[115] Griffiths, G.L., Goldenberg, D.M., Jones, A.L. and Hansen, H.J., Bioconjugate Chem., 3 (1992) 91–99.

[116] Griffiths, G.L., Goldenberg, D.M., Diril, H. and Hansen, H.J., Cancer (Suppl.), 73 (1994) 761–768.

[116a] Winnard, P. Jr., Virzi, F., Fogarasi, M., Rusckowski, M. and Hnatowich, D.J., Q. J. Nucl. Med. 40 (1996) 151–160.

[117] John, E., Thakur, M.L., DeFulvio, J., McDevitt, M.R. and Damjanov, I., J. Nucl. Med., 34 (1993) 260–267.

[118] Sykes, T.R., Woo, T.K., Baum, R.P., Qi, P., Noujaim, A.A., J. Nucl. Med., 36 (1995) 1913–1922.

[119] Mather, S.J. and Ellison, D., J. Nucl. Med., 31 (1990) 692–697.

[120] Varnum, J.M., Thakur, M.L., Schally, A.V., Jansen, S. and Mayo, K.H., J. Biol. Chem., 269 (1994) 12583–12588.
[120a] Jankowski, R., Kirsch, S., Reich, T., Spies, H. and Johannsen, B., J. Inorg. Biochem., 70 (1998) 99–106.
[121] Haberberger, T., Zamora, P., Hosono, M., Hosono, M.N., Bender, H., Guhlke, S., Knap, F.F. Jr. and Biersack, H.J., in Nicolini, M., Bandoli, G. and Mazzi, U. (eds.), *Technetium and Rhenium in Chemistry and Nuclear Medicine 4*, Proceedings of the symposium held in Bressanone, Italy, 1995, SGEditoriali, Italy 1995, pp. 367–372.
[121a] Zamora, P.O., Bender, H., Knapp, F.F. Jr., Rhodes, B.A. and Biersack, H.J., Hybridoma, 16 (1997) 85–91.
[122] Zamora, P.O., Guhlke, S., Bender, H., Diekmann, D., Rhodes, B.A., Biersack, H.J. and Knapp, F.F. Jr., Int. J. Cancer, 65 (1996) 214–220.
[123] Serafini, A.N., Int. J. Radiat. Oncol. Biol. Phys., 30 (1994) 1187–1194.
[124] Mathieu, L., Chevalier, P., Galy, G. and Berger, M., Int. J. Appl. Rad. Isot., 30 (1979) 725–727.
[125] Deutsch, E., Libson, K., Vanderheyden, J.-L., Ketring, A.R. and Maxon, H.R., Nucl. Med. Biol. 13 (1986) 465–477.
[126] De Klerk, J.M.H., Van Dijk, A., Van het Schip, A.D., Zonnenberg, B.A. and Van Rijk, P.P., J. Nucl. Med. 33 (1992) 646–651.
[127] Van het Schip, A.D., De Klerk, J.M.H., Van Dijk, A., Zonnenberg, B.A. and Van Rijk, P.P., Eur. J. Nucl. Med., 20 (1993) 876 (abstract).
[128] Francis, M.D. and Fogelman, I., in: Fogelman, I. (ed.), *Bone Scanning in Clinical Practice*, Springer-Verlag, Berlin, Heidelberg, 1987, pp. 7–17.
[129] Wieninger, J., Ketring, A.R. and Deutsch, E.A., Nucl. Med., 23 (1984) 81–2.
[130] Washburn, L.C., Biniakiewicz, D.S., Ehrhardt, G.J. and Maxon, H.R., J. Nucl. Med., 35 (1994) 244 (abstract).
[131] Ketring, A.R., Nucl. Med. Biol., 14 (1987) 223–232.
[132] Deutsch, E.A., European Patent 250,966, (1988).
[133] Maxon, H.R., Deutsch, E.D., Thomas, S.R., Libson, K., Lukes, S.J., Williams, C.C. and Ali, S., Radiology, 166 (1988) 501–507.
[134] Nosco, D.L., Tofe, A.J., Lyle, L.R., Wolfangel, R.G., Bushman, M.J., Grummon, G.D., Helling, D.E., Marmion, M.E., Miller, K.M., Pipes, D.W., Strubel, T.W. and Wester, D.W., in Nicolini, M., Bandoli, G. and Mazzi, U. (eds.), *Technetium and Rhenium in Chemistry and Nuclear Medicine 3*, Raven Press, New York, 1990, pp. 381–392.
[135] Panek-Finda, H., World Patent WO 92/00758 (1992).
[136] Wolfangel, R.B., World Patent WO 92,00759 (1992).
[137] Pipes, D., Yuan, J., Helmer, B., Deutsch, K., Elder, R.C. and Deutsch, E., J. Nucl. Med., 34 (1993) 38 (abstract).
[137a] Maxon, H.R., Schroder, L.E., Washburn, L.C., Thomas, S.R., Samaratunga, R.C., Biniakiewicz, D., Moulton, J.S., Cummings, D., Ehrhardt, G.J. and Morris, V., J. Nucl. Med., 39 (1998) 659–663.
[137b] Lin, W.-Y., Lin, C.-P., Yeh, S.-J., Hsieh, B.-T., Tsai, Z.-T., Ting, G., Yen, T.-C., Wang, S.-J., Knapp, F.F. Jr., Stabin, M.G., Eur. J. Nucl. Med. 24 (1997) 590–595.

[138] Maxon, H.R., Deutsch, E.A., Thomas, S.R., Moulton, J.S., Schroder, L.E., Libson, K.F., Soldano, M., Williams, C.C. and Cummings, D.D., J. Nucl. Med. 29 (1988) 776 (abstract).

[139] Maxon, H.R. III, Schroder, L.E., Thomas, S.R., Hertzberg, V.S., Deutsch, E.A., Scher, H.I., Samsaratunga, R., Libson, K.F., Williams, C.C., Moulton, J.S. and Schneider, H.J., Radiology, 176 (1990) 155–159.

[140] Maxon, H.R. III, Schroder, L.E., Hertzberg, V.S., Thomas, S.R., Englaro, E.E., Samaratunga, R., Smith, H., Moulton, J.S., Williams, C.C., Ehrhardt, G.J. and Schneiger, H.J., J. Nucl. Med., 32 (1991) 1877–1881.

[141] Maxon, H.R. III, Thomas, S.R., Hertzberg, V.S., Schroder, L.E., Englaro, E.E., Samsaratunga, R., Scher, H.I., Moulton, J.S., Deutsch, E.A., Deutsch, K.F., Schneider, H.J., Williams, C.C. and Ehrhardt, G.J., Sem. Nucl. Med., 22 (1992) 33–40.

[142] Maxon, H.R., Thomas, S.R., Samaratunga, R., Washburn, L.C., Schroder, L.E., Moulton, J.S., Bracken, R.B. and Vespie, A., J. Nucl. Med. 33 (1992) 937 (abstract).

[143] De Klerk, J.M.H., Zonnenberg, B.A., Quirijnen, J.M.S.P., Van het Schip, A.D., Van Dijk, A. and Van Rijk, P.P., Eur. J. Nucl. Med., 20 (1993) 876 (abstract).

[144] De Klerk, J.M.H., Zonnenberg, B.A., Van het Schip, A.D., Van Dijk, A., Han, S.H., Quirijnen, J.M.S.P., Blijham, G.H. and Van Rijk, P.P., Eur. J. Nucl. Med., 21 (1994) 1114–1120.

[145] Blower, P.J., Singh, J. and Clarke, S.E.M., J. Nucl. Med. 32 (1991) 845–849.

[146] Sampson, C., Nucl. Med. Commun., 8 (1987) 184–185.

[147] Bisunadan, M.M., Blower, P.J., Clarke, S.E.M., Singh, J. and Went, M.J., Appl. Radiat. Isot., 42 (1991) 167–171.

[148] Singh, J., Reghebi, K., Lazarus, C.R., Clarke, S.E.M., Callahan, A.P., Knapp, F.F. and Blower, P.J., Nucl. Med. Commun. 14 (1993) 197–203.

[149] Lisic, E.C., Mirzadeh, S. and Knapp, F.F. Jr., J. Labelled Compounds Radiopharm., 33 (1993) 65–75.

[150] Singh, J., Powell, A.K., Clarke, S.E.M. and Blower, P.J., J. Chem. Soc., Chem., Commun. 1115–1117.

[151] Blower, P.J., Singh, J., Clarke, S.E.M., Bisunadan, M.M. and Went, M.J., J. Nucl. Med. 31 (1990) 768 (abstract).

[152] Blower, P.J., Lam, A.S.K., O'Doherty, M.J., Kettle, A.G., Coakley, A.J., Knapp, F.F. Jr., Eur. J. Nucl. Med., 25 (1998) 613–621.

[152a] Dadachova, E. and Chapman, J., Nucl. Med. Commun., 19 (1998) 173–181.

[153] Chi, D.Y., O'Neil, J.P., Anderson, C.A., Welch, M.J. and Katzenellenbogen, J.A., J. Med. Chem., 37 (1994) 928–937.

[154] O'Neil, J.P., Wilson, S.R. and Katzenellenbogen, J.A., Inorg. Chem., 33 (1994) 319–323.

[155] Askienazy, S., Turak, B., Piketty, M.L., Munari, C., Habert, M.O., Lebtahi, R., Dilouya, A., Chodkiewicz, J.P. and Sante-Anne, C.H., Eur. J. Nucl. Med., 16:S143 (abstract).

[156] Chinol, M., Vallabhajosula, S., Goldsmith, S.J., Klein, M.J., Deutsch, K.F., Chinen, L.K., Brodack, J.W., Deutsch, E.A., Watson, B.A. and Tofe, A.J., J. Nucl. Med., 34 (1993) 1536–1542.

[156a] Wang, S.J., Lin, W.Y., Chen, M.N., Chi, C.S., Chen, J.T., Ho, W.L., Hsieh, B.T., Shen, L.H., Tsai, Z.T., Ting, G., Mirzadeh, S. and Knapp, F.F., J. Nucl. Med., 39 (1998) 1752–1757.

[156b] Wang, S.J., Lin, W.Y., Chen, M.N., Hsieh, B.T., Shen, L.H., Tsai, Z.T., Ting, G., Chen, J.T., Ho, W.L., Mirzadeh, S. and Knapp, F.F., Nucl. Med. Commun., 19 (1998) 427–433.

[157] Le Jeune, J.J., Ermias, A., Beau, P., Perdrisot, R., Pouliquen, D., Leger, J., Le Pape, A. and Jallet, P., Eur. J. Nucl. Med., 16 (1990) S143 (abstract).

[158] Hafeli, U., Tiefenauer, L.X., Schubiger, P.A. and Weder, H.G., Nucl. Med. Biol., 18 (1991) 449–454.

[159] Deutsch, E.A., US Patent 4839467 (1989).

[160] Dadachova, E., Smith, S.V., Schmidt, P.F. and Mirzadeh, S., J. Labelled Compounds Radiopharm., 37 (1995) 524–525 (abstract).

[161] Vanderheyden, J.-L., Fritzberg, A.R., Bugaj, J.E. and Venkatesan, P., World Patent WO 91/04057 (1991).

[162] Su, F.-M., Vanderheyden, J.-L., Bugaj, J.E., Venkatesan, P. and Fritzberg, A.R., J. Nucl. Med., 33 (1992) 979 (abstract).

[163] De Kraker, J., Hoefnagel, C.A., Caron, H., Valdes Olmos, R.A., Zsiros, J., Heij, H.A. and Voute, P.A., Eur. J. Cancer, 31A (1995) 600–602.

MACROCYCLIC POLYAMINES AND THEIR METAL COMPLEXES:

A NOVEL TYPE OF ANTI-HIV AGENT*

Eiichi Kimura, Tohru Koike, and Yoshio Inouye

*This review is dedicated to the memory of the late Professor R. W. Hay.

1. INTRODUCTION

While the basic and or applied chemistry of saturated macrocyclic polyamines, typically represented by 1,4,7,10-tetraazacyclododecane (cyclen, **1**) and 1,4,8,11-tetraazacyclotetradecane (cyclam, **2**), have both

Perspectives on Bioinorganic Chemistry
Volume 4, pages 145–164.

ISBN: 0-7623-0352-2

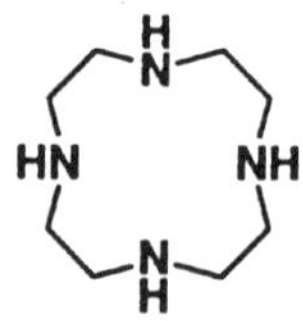

1
cyclen

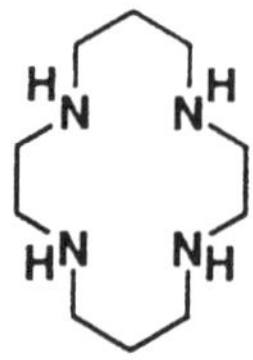

2
cyclam

been well developed [1–3], their biochemical and medicinal potentials have not been fully exploited [4–8].

In March 1992 [9], Kimura's group reported that metal complexes of macrocyclic polyamines such as **1** and **2**, although slightly, selectively inhibited replication of several strains (IIIb, RF, WN) of human immunodeficiency virus type 1 (HIV-1), while their toxicities to the host MT-4 cell (a human T lymphocyte) were weaker. The free ligands **1** and **2** also showed anti-HIV activity, but their cytotoxicities were too high to permit accurate measurement of their anti-HIV activity. Independently, in June 1992 De Clercq's group published a paper [10] whereby cyclam **2** and a series of biscyclam derivatives (e.g. **3**) were shown to have strong activity against HIV-1 (IIIb, RF, HE, etc.) and human immunodeficiency virus type 2 (HIV-2) (ROD, EHO, etc.). In particular, biscyclam **3** was much more potent at an effective concentration as low as $EC_{50} = 0.14$ μM than monomeric cyclam **2** ($EC_{50} = 399$ μM) against HIV-1_{IIIb}, although the cytotoxicity towards host MT-4 cells remained fairly high. Moreover, they found that biscyclam **3** was active against 3′-azido-3′-deoxythimidine (AZT)-resistant HIV-strains and acted additively with AZT. The HIV-inhibition mechanism seemed to occur at an early event of the retrovirus replication cycle (see Figure 1), tentatively identified as a viral uncoating event.

Following these independent discoveries by the two groups, a number of macrocyclic polyamine derivatives have been assayed and synthesized

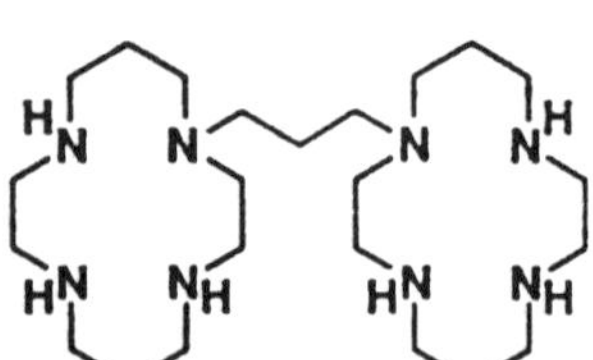

3

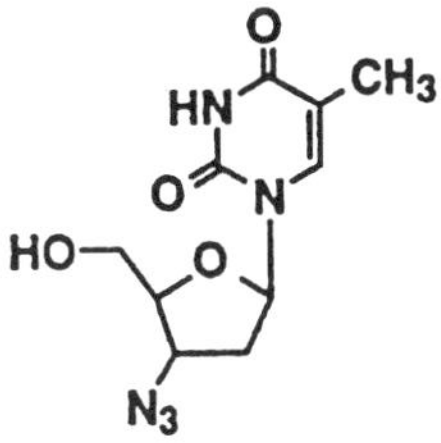

AZT

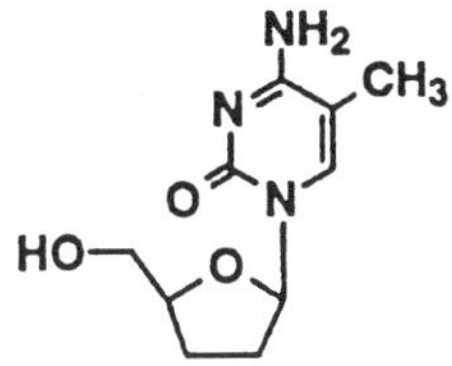

DDC

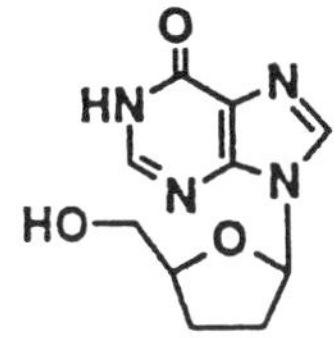

DDI

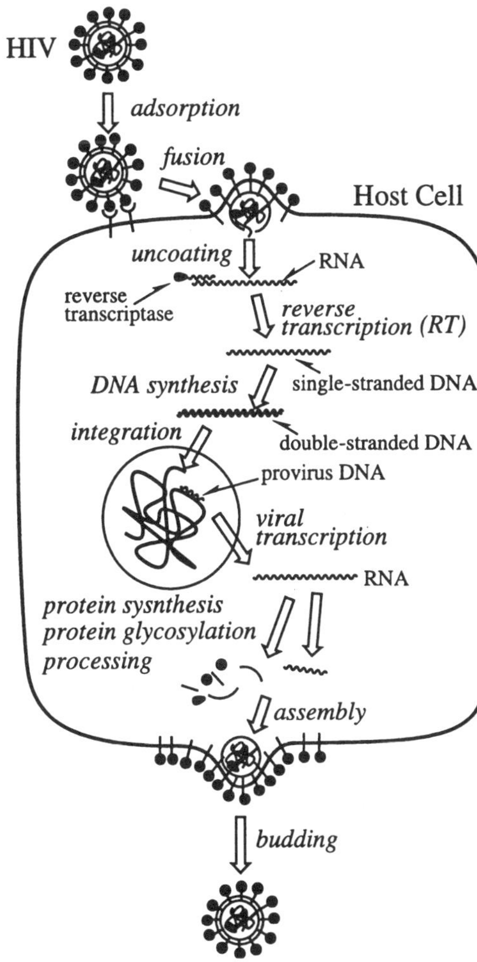

Figure 1. Replication cycle of human immunodeficiency virus (HIV).

to seek more potent and less toxic drug candidates. Among the current chemotherapeutic compounds for HIV, the most extensively studied are terminators of DNA synthesis during the reverse transcription (RT) reaction (e.g. AZT, DDC, and DDI), and inhibitors of HIV protease, an essential proteolytic enzyme required for the assembly of fully infectious viral particles [11]. Although macrocyclic polyamines and their metal complexes were previously unknown as HIV-active compounds, the fact that (1) their HIV-inhibition mechanism seemed new, (2) their cytotoxicity against uninfected host cells seemed manageable, (3) they were active against AZT-resistant strains, and (4) their chemistry was already extremely well understood, suggests that these compounds offer an appropriate prototype for potential new anti-HIV drugs. This chapter reviews the current state of development of these compounds as anti-HIV agents.

2. MACROCYCLIC POLYAMINE COMPOUNDS

Typical macrocyclic polyamine compounds (**4–25**) tested by Kimura's group [12–14] are shown in Figure 2 and those (**26–35**) by De Clercq's group [10,15,16] in Figure 3.

3. ANTI-HIV ACTIVITIES OF MACROCYCLIC POLYAMINE COMPOUNDS

Kimura'group [12,13] and De Clercq's group [10,15,16] employed a similar, well-established method for antiviral activity assays and cytotoxicity assays. The anti-HIV activity and cytotoxicity measurements in MT-4 cells [which are highly susceptible to cytopathogenic effect (CPE) of HIV] were based on viability of the host MT-4 cells that had been or had not been infected with HIV and then exposed to various concentrations of the test compounds [17]. After the MT-4 cells were allowed to proliferate for 4–6 days, the number of viable cell was quantified by appearance of a visible absorption at 595 nm in its presence of 3-(4,5-dimethylthiazol-2-yl)-2,5-diphenyltetrazolium bromide (MTT method), which measures the reducing ability of mitochondria in living cells. The 50% effective concentration (EC_{50}) was defined as the concentration at which 50% of CPE was inhibited. The cytotoxicity of each compound was evaluated in parallel with the determination of anti-HIV activity. At the 50% cytotoxic concentration (CC_{50}), the cell viability of HIV-uninfected MT-4 cells was half that of the test compound-untreated cells. Figure 4 illustrates a typical viability curve at various concentrations of anti-HIV agent, from which the EC_{50} and CC_{50} values were evaluated. Clearly the test compounds should have $CC_{50} >> EC_{50}$ for their anti-HIV activity to be measured. The selectivity index (SI) corresponding to the ratio CC_{50}/EC_{50} (should be >1) is a measure of the efficiency as a drug (i.e. the greater a compound's SI, the greater its potential use as a drug). The observed anti-HIV activities are summarized in Table 1 for Kimura's group and in Table 2 for De Clercq's group.

It may be difficult to directly compare Kimura's EC_{50} with De Clercq's EC_{50} values for the same or relevant compounds. Kimura's group first discovered that when cyclen **1** was complexed with zinc(II) ion (**4**), the cytotoxicity of **1** decreased (i.e. CC_{50} increased) drastically by a factor of ca. 300 (see Table 1), so as to permit determination of EC_{50} [12]. The free cyclen **1** normally present as a diprotonated species at physiological pH (pK_a values of 11.0, 9.86, <2, and <2 at 25 °C [18]) failed to protect MT-4 cells from CEP of HIV-1 at the nontoxic concentrations (i.e. the EC_{50} was larger than CC_{50} value). The anti-HIV activity became apparent with its zinc(II) complex **4** owing to the decreased cytotoxicity with SI of 12. Since the metal complex served to reduce the cytotoxicity, the measurement of anti-HIV activity of cyclam **2** with various transition metal ions (e.g. Ni^{2+}, Zn^{2+}, Cu^{2+}, etc.) became possible. Although Kimura (EC_{50} = 420 μM and CC_{50} = 953 μM after 4 days) and De Clercq (EC_{50} = 399 μM and CC_{50} = 1248 μM after 5 days) both discovered cyclam **2** to possess similar anti-HIV-1 activity, Kimura found that when the cell viability was measured 6 days after its addition, its cytotoxicity significantly increased (CC_{50} = 68 μM) against the host MT-4 cells, which hindered them from observing the anti-HIV activity (see Table 1). De Clercq reported only the effects after 5 days (see Table 2). Some inconsistency between the two groups' results might be explained by the time-dependent cytotoxicity of macrocyclic polyamines. While keeping the anti-HIV-1 activity at the same or at a higher level, the delayed cytotoxicity of **2** disappeared in the Ni^{2+}, Zn^{2+}, and Cu^{2+} complexes **5**, **6**, and **7**, respectively. While these divalent metal complexes maintained the inhibition of the replication of HIV-1, the trivalent Fe^{3+} and Co^{3+} complexes with cyclam failed to show the anti-HIV activity. Cyclams with a pendant donor atom either as metal-free ligands or metal complexes showed strong cytotoxicity. Thus, the tetracoordinated complex structure (either planar or tetragonal) seems optimal for the interaction with the target HIV-receptor site. The results of our experiments [13] (see Table 1), on other monocyclic compounds such as triamine [12]aneN_3 **15**, pentaamine [15]aneN_5 **19**, and hexaamine [18]aneN_6 **21** showed that the

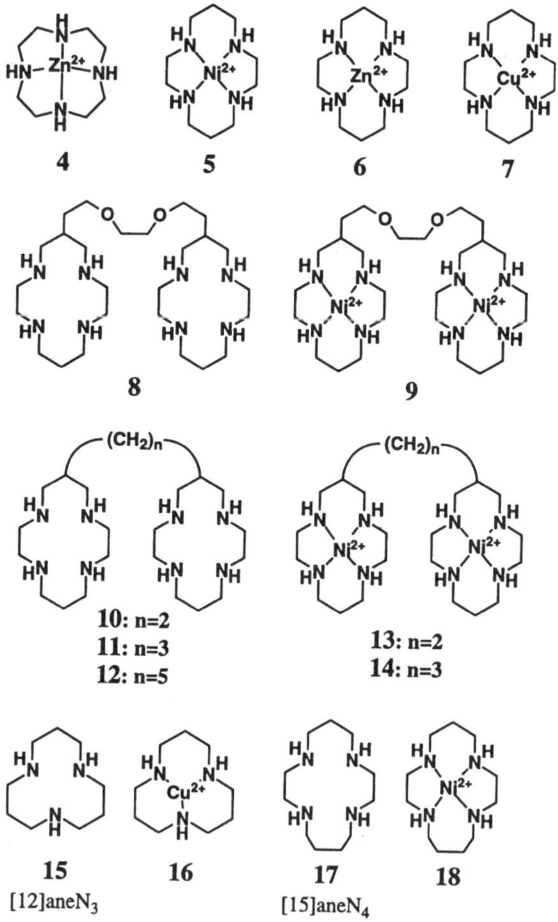

Figure 2. Structures of typical macrocyclic polyamine compounds studied by Kimura's group.

high cytotoxicity of these macrocycles precluded the determination of anti-HIV activity. The only exception was a 15-membered tetraamine [15]aneN_4 **17** that showed stronger inhibition of HIV proliferation and weaker cytotoxicity against MT-4 cells than those of the 14-membered cyclam **2**.

A dimeric cyclam (biscyclam) **8** [12], with two cyclam moieties linked by a 3,6-dioxaoctane-1,8-diyl group at the C(6) positions, showed a remarkably improved anti-HIV activity after 4 days with the EC_{50} value

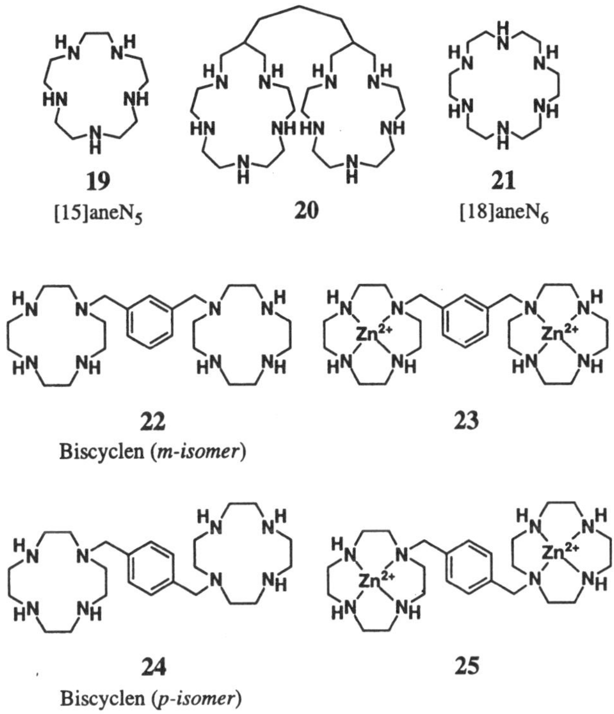

Figure 2. Continued

being ca. 1/130 reduced with respect to that of cyclam **2**. However, as was the case with **2**, the cytotoxicity of **8** increased significantly from 4 days after its addition (CC_{50} = 342 μM) to 6 days (CC_{50} = 27 μM), which resulted in a 30-fold decrease in SI. Its dinuclear nickel(II) complex **9** possessed almost twice the anti-HIV activity, but also twice the cytotoxicity of the parent dimer ligand **8**. There was little delayed cytotoxicity with the nickel(II) complex **9**.

De Clercq's biscyclams, especially **3** and **26** [10], showed extremely improved anti-HIV-1 activity EC_{50} of 0.248 and 0.144 μM, respectively, resulting in very high SI values of >2500 and 2215 with respect to cyclam **2** (SI = 3) (see Table 2). These dimeric cyclams **3** and **26** were also effective against other strains of HIV-1 and HIV-2 (e.g. HIV-1_{RF}, HIV-2_{ROD}). Irrespective of the method used to assay anti-HIV activity (i.e. inhibition of viral cytopathicity, antigen expression, etc.), the biscyclams were invariably active against HIV-1 and HIV-2. Very interestingly, other

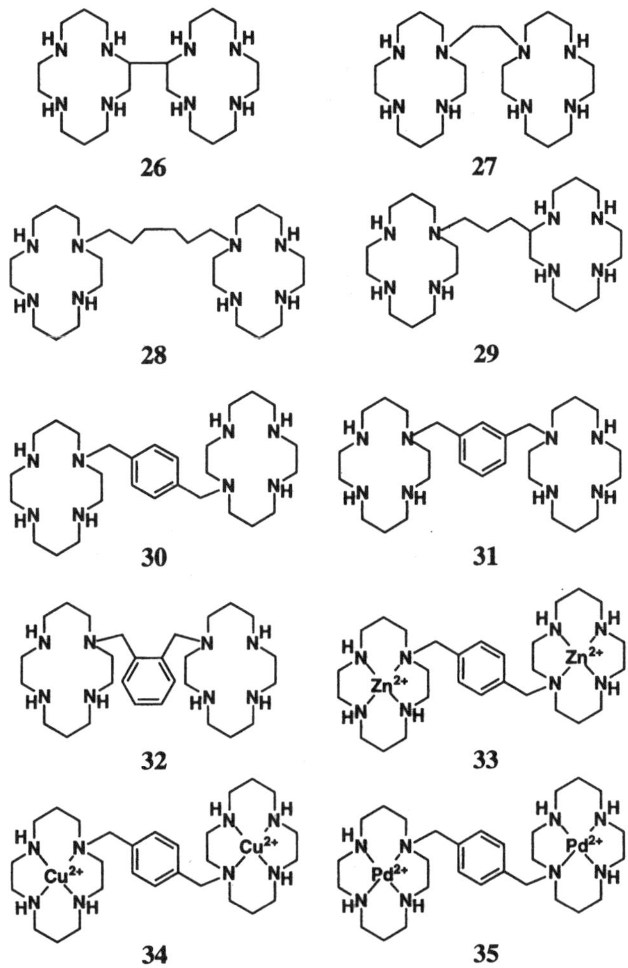

Figure 3. Structures of typical macrocyclic polyamine compounds studied by De Clercq's group.

viruses (e.g. influenza, measles, herpes) were all insensitive to **3** and **26** at concentrations up to ca. 500 μM. Compound **3** was nontoxic to the host cells (MT-4) at concentrations up to 622 μM. When **3** was assayed against HIV-1 in combination with AZT, the compounds appeared to act in an additive, but not in a synergistic or antagonistic fashion. De Clercq's group concluded in their first publication [10] that **3** seems to interact with an early event of the retrovirus replication cycle, possibly at viral

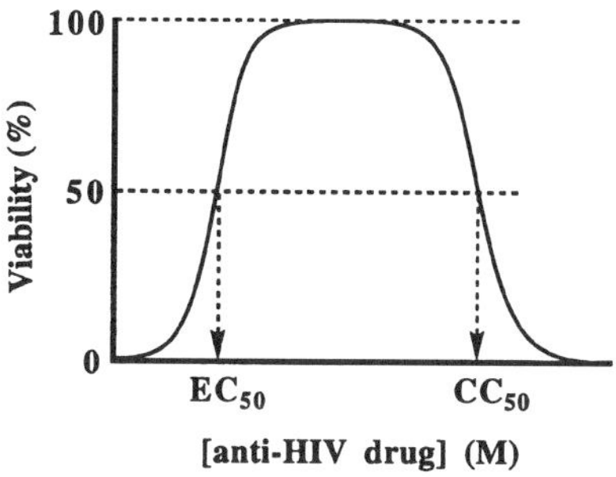

Figure 4. A typical viability curve for HIV-infected cell in the presence of anti-HIVdrug. EC50 is 50% antiviral effective concentration. CC50 is 50% cytotoxic concentration.

uncoating event. If so, the biscyclams represent the first compounds with this unique mode of action.

Kimura's group further tested their biscyclams [13]. Among the *C,C′*-linked biscyclams and their nickel(II) complexes (**8–14**), **11** exhibited the highest anti-HIV-1_{IIIb} activity (EC_{50} = 0.08 μM) and very low cytotoxicity (CC_{50} = 400 μM), resulting in an extremely large SI of 5000 after 4 days of incubation. Its nickel(II) complex **14** appreciably reduced the anti-HIV activity (EC_{50} = 0.47 μM), while the cytotoxicity was not so affected (CC_{50} = 200 μM). Later, the Kimura group found that dimeric cyclens (biscyclens) **22** and **24** with a xylyl linker are also extremely promising anti-HIV-1_{IIIb} agents (i.e. EC_{50} = 0.088 and 0.26 μM, respectively) [14]. Most notably, when the *m*-xylyl biscyclen **22** was complexed with two zinc(II) ions to form **23**, the cytotoxicity was drastically reduced (CC_{50} >1100 μM). Thus, while the metal-free parent compound **22** has SI = 70, its dinuclear zinc(II) complex **23** has an extremely large SI value of >22,000. By comparison, against HIV-1_{IIIb} AZT showed EC_{50} = 0.0082 μM, CC_{50} = 120 μM, and SI = 15,000 after 4 days by the same assay method (see Table 1). Analogous compounds DDI and DDC showed EC_{50} of 18 and 6.6 μM, and CC_{50} of 210 and 40 μM, respectively. Likewise dextran sulfate 8000 which is also an effective anti-HIV agent showed EC_{50} = 0.21 μM, CC_{50} >100 μM, and SI >480. Accordingly, the *in vitro* test by the Kimura group also suggested the feasibility of bis(macrocyclic tetraamines) as attractive candidate drugs for the chemotherapy and prophylaxis of HIV-1 infections.

Table 1. Anti-HIV-1_{IIIb} Activities of Macrocyclic Compounds (Kimura's Group) in MT-4 Cells

Compound	*Days*	*EC_{50} (μM)*	*CC_{50} (μM)*	*SI*
1, cyclen	4 (6)	NA (NA)[a]	20 (14)	
4, Zn^{2+}-cyclen	4 (6)	515 (370)	6270 (4530)	12 (12)
2, cyclam	4 (6)	420 (NA)	953 (68)	2
5, Ni^{2+}-cyclam	4 (6)	120 (271)	1610 (2140)	13 (8)
6, Zn^{2+}-cyclam	4 (6)	22 (36)	450 (448)	20 (13)
7, Cu^{2+}-cyclam	4 (6)	120 (460)	2520 (3220)	21 (7)
8	4 (6)	3.2 (9.3)	342 (27)	107 (3)
9	4 (6)	1.9 (2.5)	170 (201)	91 (81)
10	4 (5)	0.13 (0.18)	360 (50)	2800 (280)
11	4 (5)	0.08 (0.12)	400 (300)	5000 (2500)
12	4 (5)	0.29 (0.67)	210 (110)	720 (160)
13	4 (5)	0.37 (0.48)	200 (500)	540 (1100)
14	4 (5)	0.47 (0.30)	200 (640)	430 (2100)
15	4	NA	85	
16	4	NA	15	
17	4	100	>930	>9
18	4	49	100	2
19	4	NA	8.3	
20	5	NA	4.5	
21	4	NA	<8	
22	5	0.088	6.2	70
23	5	0.049	>1100	>22000
24	5	0.26	13	50
25	5	0.14	750	5400
AZT	4 (5)	0.0082 (0.0075)	120 (119)	15000 (14500)
dextran sulfate 8000	6	0.21	>100	>470
$K_7[PTi_2W_{10}O_{40}]\cdot 6H_2O$	6	1.36	92	68

Note: [a]Not available because CEP inhibition was less than 50% at nontoxic concentration.

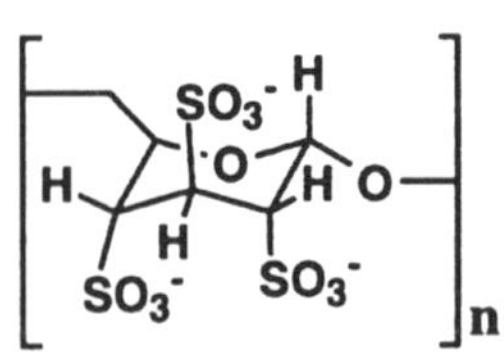

dextran sulfate

Table 2. Anti-HIV-1_{IIIb} Activities of Macrocyclic Compounds (De Clercq's Group) in MT-4 Cells

Compound	*Days*	*EC_{50} (μM)*	*CC_{50} (μM)*	*SI*
2, cyclam	5	399	1248	3
3	5	0.248	>622	>2500
22	5	0.075	20	267
24	5	0.322	55	171
26	5	0.144	319	2215
27	5	18.6	349	19
28	5	0.616	290	470
29	5	1.96	>445	>225
30	5	0.004	>421	>100000
31	5	0.034	>421	>12000
32	5	1.357	>168	>120
33	5	0.003	>251	>76000
34	5	0.018	>201	>11000
35	5	32	>210	>6.7

Following the initial biscyclam with alkyl linkers, De Clercq's group synthesized a large number of biscyclams to find that among biscyclens (**22** and **24**), biscyclam derivatives linked with xylyl linkers (**30**, **31**, and **32**), and quite a number of homologues (dimeric 12–16-membered macrocyclic tetraamines), the *p*-xylyl biscyclam **30** had the most potent anti-HIV-1_{IIIb} activity (EC_{50} = 0.004 μM) and very low MT-4 cell cytotoxicity (CC_{50} > 421 μM), and anti-HIV-2_{ROD} activity (EC_{50} = 0.0059 μM) [19]. The metal complexation effect on **30** was also investigated with zinc(II) (**33**), copper(II) (**34**), and palladium(II) complexes (**35**). Metal complexes of low kinetic stability such as dizinc(II) complex **33** retained activity comparable to that of the parent compound. From the more detailed study with variously substituted aromatic linkers, De Clercq's group concluded that the activity of biscyclams appears to be insensitive to electron-withdrawing or electron-donating properties of substituents onto the linker, while sterically hindering linkers such as biphenyl group markedly reduced activity. As a result, several analogues with anti-HIV potency comparable to that of **30** have been identified. However, **30** remained the most active congener. The quantitative structural activity relationship (QSAR) study was then reported [16,20].

4. MODE OF ANTI-HIV ACTION BY MACROCYCLIC POLYAMINE COMPOUNDS

To determine at which stage the biscyclams **26** and **3** actually interact with the HIV replicative cycle (see Figure 1), De Clercq's group initially conducted a time-of-addition experiment [10]. The cells were infected at high virus multiplicity to ensure that the virus replicative steps would be synchronized in the whole cell population, and the compounds were added at 0, 1, 2, . . . or 24 (n) h after infection. Depending on the stage at which they interact and the need for intracellular metabolism, addition of the compounds could be delayed for n h without loss of activity. Dextran sulfate, which acts at the virus adsorption step, must be added together with the virus ($n = 0$) to be active. In the case of AZT and DDI, which following intracellular phosphorylation act at the reverse transcription (*RT*) step, addition to the cell could be delayed until $n =$ ca. 5 h after infection. The protease inhibitor Ro31-8959, which interacts with a late event in the virus cycle (assembly of mature virions), was still effective if added as late as $n = 21$ h after infection. From the time-of-addition experiment, De Clercq proposed that the biscyclams ($n = 1$ or 2 h) had to interact with a process following virus adsorption but preceding reverse transcription, which means virus–cell fusion and/or uncoating.

When uninfected CD4-positive cells (having CD4 receptor) and HIV-infected cells (with gp120 at the surface) are cocultured, the multinucleated giant cells are formed due to the cell–cell fusion (syncytium formation) [21] (see Figure 5). Since the mechanism of cell–cell fusion is considered to be closely related to that of HIV-cell fusion, the syncytium formation has often been used as a surrogate measure of HIV-cell fusion. To know more about the mode of action of biscyclams on the HIV infection, De Clercq's group [19] and Kimura's group [12] investigated the effects of macrocyclic polyamines on syncytium formation. According to Kimura's group, the biscyclams were added to the

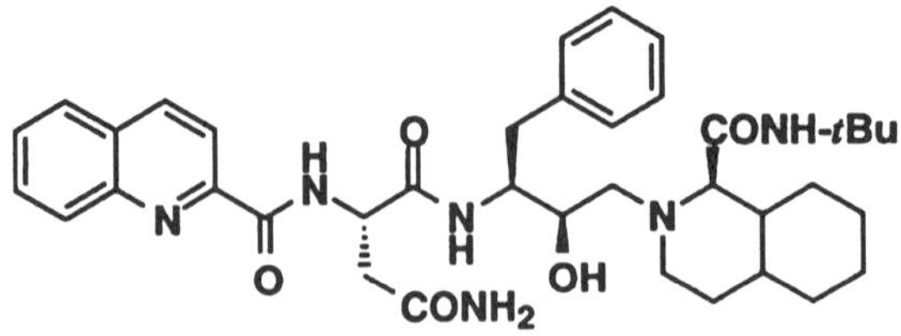

Ro31-8959

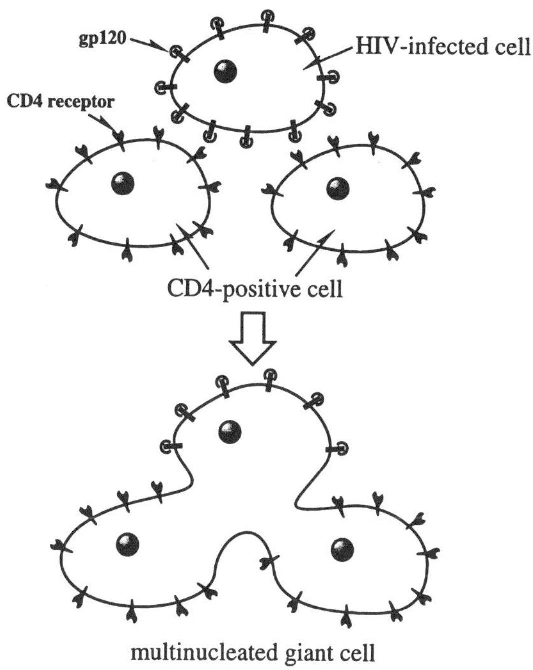

Figure 5. Formation of multinucleated giant cell from HIV-infected cell and CD4-positive cell.

cocultures of uninfected and HIV-infected MOLT-4 (MOLT-4/HIV) cells; both cells were cocultured at the final cell density of 2.5×10^5 cells/mL in a mixture of 2:1 or individually in the presence of various concentrations of test compounds at 37 °C for 20 h. Control wells received either MOLT-4 cells, MOLT-4/HIV cells, or a 2:1 mixture in the absence of test compounds at the same cell density as added with the compounds. The fusion index (FI) was defined as follows: FI = [(cell number in MOLT-4 well) × 2 + (cell number in MOLT-4/HIV well)]/[(cell number in mixed-culture well) × 3] – 1.0. Generally the FIs in the control cultures ranged from 0.5 to 1.2. Percent reduction in FI = (1 – FIT/FIC) × 100, where FIT is FI for the test compound and FIC is that of the control. The 50% inhibitory concentration (IC_{50}) was defined as the concentration at which FI was reduced by 50%.

De Clercq's group did not observe any inhibition of syncytium formation by the biscyclam containing an aliphatic linker (**3**), but observed it in the case of the biscyclam containing an aromatic linker (**30**), albeit at three orders of higher concentration than needed for the inhibition of viral CPE [19]. A linear activity correlation between inhibition of HIV-1 (HIV-2) and the syncytium formation inhibition by biscyclams containing aromatic linkers suggested a common target, namely gp120, which is involved at the fusion with host cell membrane. Kimura's group observed almost comparable activity for inhibition of HIV-1_{IIIb}-induced syncytium formation and the inhibition of HIV-induced CPE by their biscyclams (**8–14**) and biscyclens (**22–25**) in both parent ligand and metal complex forms [12]. It was also established that the anti-HIV compounds such as polyoxometalate $K_7[PTi_2W_{10}O_{40}]\cdot 6H_2O$ (IC_{50} = 5.1 μM) and dextran sulfates 8000 (IC_{50} = 0.58 μM) inhibit the HIV-1_{IIIb}-induced syncytium formation, suggesting the activity at a similar process. By contrast, AZT did not show this activity (IC_{50} >186 μM).

A currently accepted concept of early events in HIV infection is as follows. The interaction between nonlinear epitope of gp120 and one site on the CD4 molecule (i.e. the CDR2 domain) of the host T cells, if strengthened by a secondary interaction between a certain portion of gp120 (most probably the V3 loop, see Figure 6) and recently identified cellular cofactors (or co-receptors) [22–24], permits the interaction of the hydrophobic fusion domain on gp41 with the target cell membrane (see Figure 7). While the V3 loop of HIV-1 and the corresponding region of HIV-2 (the V3-like loop) are poorly conserved in mutation, they share a common nature of high local positive charge density with arginine and

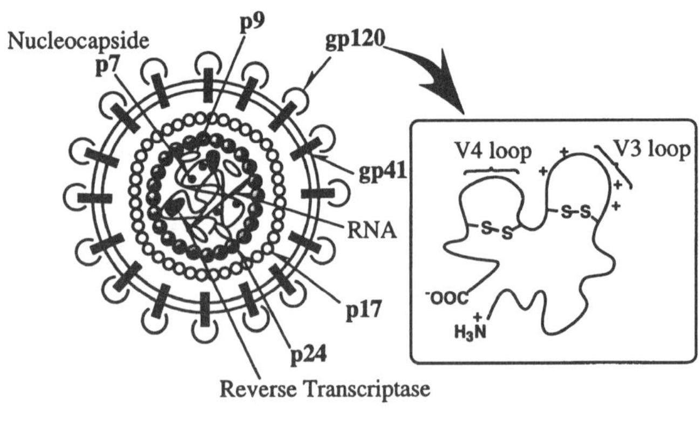

Figure 6. Structure of a HIV vilion with envelope glycoproteins and nucleocapside proteins. Magnified wire structure of its glysoprotein gp120 shows V3 and V4 loops with S-S bonding sites.

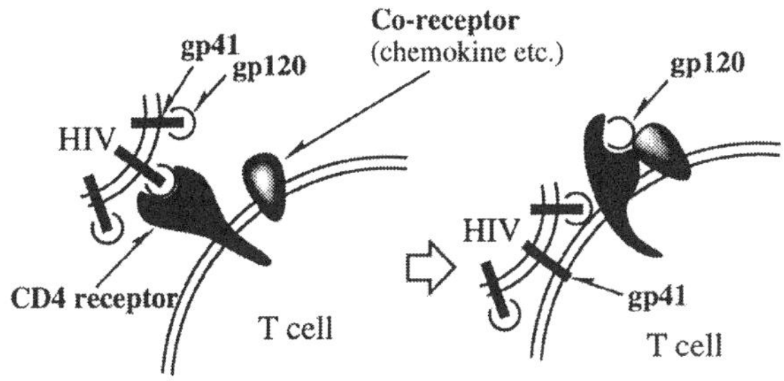

Figure 7. A recently proposed HIV adsorption/fusion mechanism onto T cell's CD4 receptor with coreceptor (chemokine, etc.), followed by breaching the T-cell membrane by HIV glycoprotein gp41.

lysine residues [11,21]. Polyanionic compounds (e.g. dextran sulfate 8000) inhibit HIV infection and syncytium formation by ionic interaction with the V3 loop [25].

It once was suspected that nucleocapsid (NC) protein p7 may be a possible target for biscyclams. NC p7 contains two zinc fingers [26] (see Figures 6 and 8), which participate in several nucleic acid interactions such as specific recognition of the viral RNA genome during budding, packaging of RNA in mature virions, and initiation of reverse transcription [27]. Aromatic C-nitroso compounds such as 3-nitrosobenzamide and 6-nitroso-1,2-benzopyrone have been shown to inhibit infection of HIV-2 in human lymphocytes by extracting zinc(II) ion via an oxidative

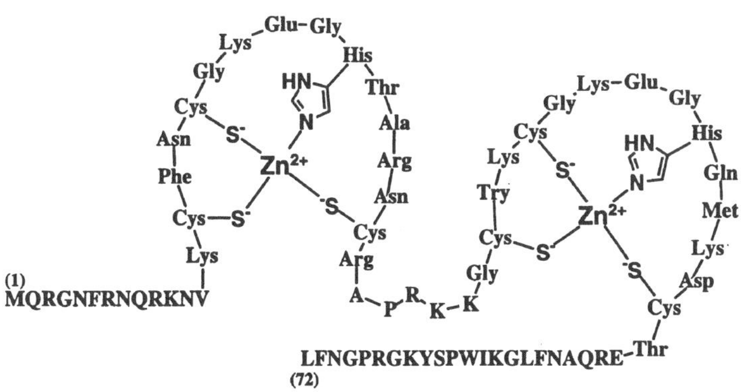

Figure 8. Primary structure of the nucleocapsid protein p7 with the two zinc fingers.

mechanism from the zinc fingers of NC p7 [28]. If extraction of zinc(II) from the zinc fingers by the bis-macrocycles occurred, this would result in a similar inhibition profile to the C-nitroso compounds. However, Joao et al. did not observe that biscyclam **30** binds or extracts zinc(II) from the double zinc finger peptide of NC p7 under the conditions where 3-nitrosobenzamide was observed to extract zinc [16]. It was thus concluded that the activity of the biscyclam probably was not due to an interaction with NC p7.

Recently, De Clercq's group showed that a monoclonal antibody which binds to the V3 loop of native gp120 is no longer able to bind to the compound **30**-resistant mutant gp120 [29]. Evidence supporting the suggestion of gp120 as a target for the bis(macrocyclic tetraamine) comes from analysis of mutant virus, resistant to biscyclam **30**, in which the mutations occurred in gp120, predominantly in the V3 loop, and not NC p7. None of the amino acids mutated in gp120 of the biscyclam-resistant strains were candidate binding partners for biscyclams. Further, it was found that the number of positively charged amino acids (i.e. arginine and lysine residues) in the V3 loop increased in the V3 loop of the biscyclam-resistant strains, implying that the positively charged biscyclams do not directly bind to the native V3 loop.

The binding sites for biscyclams may be carboxylate or sulfide anions. Since the cyclams are in diprotonated forms at physiological pH (cf. cyclam's pK_a values = 11.5, 10.2, <2, and <2 at 25 °C [3]), this postulate is renderable. In fact, bis(macrocyclic polyamine) **8**, **10**, **20**, **22**, and **24** [12–14] were originally designed by Kimura et al. as host molecules for polyanions (e.g. polycarboxylates) in aqueous solution at neutral pH [30–35]. The mutations close to the disulfide bridges of the V3 and V4 loops (see Figure 6) may indicate the cysteines (reduced species of the S–S bonds) as potential binding sites [29]. Recently, inhibitors of protein disulfide isomerases have been found to inhibit an early, postbinding step of HIV replication [36]. It was proposed that the reduction of disulfide bridges by a membrane-bound protein disulfide isomerase may be the first and necessary step in a cascade of conformational changes in gp120 initiated after binding to CD4 receptor of cells. These events should ultimately lead to an unfolding of the hydrophobic N-terminal extracellular domain in gp41, which is thought to mediate fusion of the viral membrane with the host cell membrane to initiate entry of the viral capsid. Therefore, De Clercq et al. suggested that biscyclams interfere with fusion and entry of HIV by interacting with sulfide anions of the V3 and V4 loops that are formed upon reduction of the disulfide bridges during the enzyme-catalyzed isomerization. It is to be added that the cysteines, and consequently the disulfide bridges in gp120, are highly conserved among all HIV strains, and their mutation has been shown to affect assembly and infectivity of the virus [29]. From a chemical point of view, however, it is yet to be proven that highly protonated macrocyclic polyamines interact with sulfide anions.

To determine if the high *in vitro* potents of the anti-HIV compound **30** translates into antiviral efficiency *in vivo*, Datema et al. investigated the inhibition of HIV-1 production and of depletion of human T cells in HIV-1-infected SCID-hu Thy/Liv mice [37]. Steady levels of 100 ng of **30** or higher per mL in plasma resulted in significant inhibition of HIV p24 protein formation. Daily injection of **30** caused a dose-dependent decrease in viral p24 production, and this inhibition could be potentiated by coadministration of AZT (or DDI). This study suggested that **30** alone or in combination with the licensed anti-HIV agents AZT and DDI may decrease the virus load in HIV-infected patients and, by extension, that the infectious cell entry step is a valid target for antiviral chemotherapy of HIV disease.

When administered subcutaneously in an animal model using rabbits, sufficient serum level of **30** was observed and it is extensively bound to serum proteins but remained antiviral active. A method based on the UV absorption of its copper(II) complex was used to analyze the drug concentrations in serum by HPLC. Being active even in a serum protein-bound form, **30** must act on its molecular target extracellularly, in accordance with the protruded allocation of gp120 out of membrane. In contrast, **30** could not be detected in rabbit serum after oral administration. Having a charge of 4+ under physiological conditions, it seems unlikely that **30** enters cells rapidly. The same might be responsible for their poor oral bioavailability.

5. SUMMARY AND PERSPECTIVES

So far the most potent anti-HIV compounds *in vitro* among the macrocyclic polyamines are dinuclear zinc(II) complex of *m*-xylyl-biscyclen (**23**) (by Kimura's group) and *p*-xylyl-biscyclam (**30**) (by De Clercq's group). They are active against various strains of HIV-1 and HIV-2, while cytotoxicity towards host MT-4 cells are minimal. Their activity against HIV is selective and they are not active against other viruses tested. Hence, the study of their mode of action is not only useful in developing new anti-HIV drugs, but also in revealing the distinctive characteristics of HIV from those of other viruses.

In Kimura's study, zinc(II) complexes serve to reduce the cytotoxicity to the host cell, while retaining similar anti-HIV activity, which implies that the receptor sites are different for anti-HIV action and cytotoxic action. The bis(macrocyclic tetraamine) compounds can be tremendously strong sequestering agents for metal ions such as copper(II) and zinc(II) ions. Although De Clercq's biochemical study seems to reject a hypothesis of deprivation of zinc(II) from NC p9 for the mode of action, it still remains a possibility. It is also possible that the bis(macrocyclic tetraamine) compounds may pick up zinc(II) ions at prior stage and the resulting complexes such as **23** bind irreversibly to the cysteines at the critical neck of V3 loop of gp120. As supporting evidence, Kimura's group earlier showed extremely strong binding of the anionic thiolate group to zinc(II)-cyclen complex **4** [38].

The discovery of *in vitro* anti-HIV activity of macrocyclic polyamines by the Kimura and De Clercq groups led to development of a new type of effective anti-HIV agents. The results from the studies of the pharmacokinetics and their *in vivo* efficacy in a SCID-hu Thy/Liv mice are further encouraging indicators of their potential usefulness as new AIDS curing drugs.

ACKNOWLEDGMENTS

Kimura is thankful to the Grant-in Aid for Priority Project "Biometallics" (No. 08249103) from the Ministry of Education, Science and Culture in Japan.

NOTE ADDED IN PROOF

Recent findings on anti-HIV activity of macrocyclic polyamines are reported by Dessolin et al. [39].

REFERENCES

[1] Kimura, E., in Hiraoka, M. (ed.), *Crown Ethers and Analogous Compounds*, Elsevier Science Publishers, Amsterdam, 1992, p. 381.
[2] Kimura, E., Tetrahedron, 48 (1992) 381.
[3] Kimura, E., J. Coord. Chem., 15 (1986) 1.
[4] Kimura, E., Sakonaka, A. and Nakamoto, M., Biochim. Biophys. Acta, 674 (1981) 172.
[5] Kimura, E., Watanabe, A. and Nihira, H., Chem. Pharm. Bull., 31 (1983) 3264.

[6] Takenouchi, K., Watanabe, K., Kato, Y., Koike, T. and Kimura, E., J. Org. Chem., 58 (1993) 1955.

[7] Kimura, E., Sasada, M., Shionoya, M., Koike, T., Kurosaki, M. and Shiro, M., J. Biol. Inorg. Chem., 2 (1997) 74.

[8] Riley, D.P. and Weiss R.H., J. Am. Chem. Soc., 116 (1994) 387.

[9] Inouye, Y., Kimura, E. et al., *The 112th Annual Meeting of the Pharmaceutical Society of Japan*, Fukuoka, March 1992.

[10] De Clercq, E., Yamamoto, N., Pauwels, R., Baba, M., Schols, D., Nakashima, H., Balzarini, J., Debyser, Z., Murrer, B.A., Schwartz, D., Thornton, D., Bridger, G., Fricker, S., Henson, G., Abrams M. and Picker, D., Proc. Natl. Acad. Sci. USA, 89 (1992) 5286.

[11] Review article for anti-HIV drugs: De Clercq, E., J. Med. Chem., 38 (1995) 2491.

[12] Inouye, T., Kanamori, T., Yoshida, T., Bu, X., Shionoya, M., Koike, T. and Kimura, E., Biol. Pharm. Bull., 17 (1994) 243.

[13] Inouye, Y., Kanamori, T., Sugiyama, M., Yoshida, T., Koike, T., Shionoya, M., Enomoto, K., Suehiro, K. and Kimura, E., Antiviral Chem. Chemother., 6 (1995) 337.

[14] Inouye, Y., Kanamori, K., Yoshida, T., Koike, T., Shionoya, M., Fujioka, H. and Kimura, E., Biol. Pharm. Bull., 19 (1996) 456.

[15] Bridger, G.J., Skerlj, R.T., Thornton, D., Padmanabhan, S., Martellucci, S. A., Henson, G.W., Abrams, M.J., Yamamoto, N., De Vreese, K., Pauwels, R. and De Clercq, E., J. Med. Chem., 38 (1995) 366.

[16] Joao, H.C., De Vreese, K., Pauwels, R., De Clercq, E., Henson, G. W. and Bridger, G.J., J. Med. Chem., 38 (1995) 3865.

[17] Pauwels, R., Balzarini, J., Baba, M., Snoeck, R., Schols, D., Herdewijin, P., Desmyter, J. and De Clercq, E., J. Virol. Methods, 20 (1988) 309.

[18] Koike, T., Kajitani, S., Nakamura, I., Kimura, E. and Shiro, M., J. Am. Chem. Soc., 117 (1995) 1210.

[19] De Clercq, E., Yamamoto, N., Pauwels, R., Balzarini, J., Witvrouw, M., De Vreese, K., Debyser, Z., Rosenwirth, B., Peichl, P., Datema, R., Thornton, D., Skerlj, R., Gaul, F., Padmanabhan, S., Bridger, G., Henson, G. and Abrams, M., Antimicrob. Agents Chemother., 38 (1994) 668.

[20] Bridger, G.J., Skerlj, R.T., Padmanabhan, S., Martellucci, S.A., Henson, G.W., Abrams, M.J., Joao, H.C., Witvrouw, W., De Vreese, K., Pauwels, R. and De Clercq, E., J. Med. Chem., 39 (1996) 109.

[21] Levy, J.A., Microbiol. Reviews, 57 (1993) 183.

[22] Feng, Y., Broder, C.C., Kennedy, P.E. and Berger, E.A., Science, 272 (1996) 872.

[23] Alkhatib, G., Combadiere, C., Broder, C.C., Feng, Y., Kennedy, P. E., Murphy, P.M. and Berger, E.A., Science, 272 (1996) 1955.

[24] Cohen, J., Science, 275 (1997) 1261.

[25] Callahan, L.N., Phelan, M., Mallinson, M. and Norcross, M.A., J. Virology, 65 (1991) 1543.

[26] Morellet, N., Jullian, N., De Rocquigny, H., Maigret, B., Darlix, J.-L. and Roques, B.P., EMBO J., 11 (1992) 3059.

[27] Dannull, J., Surovoy, A., Jung, G. and Moelling, K., EMBO J., 13 (1994) 1525.

[28] Rice, W.G., Schaeffer, C.A., Harten, C.A., Villinger, F., South, T.L., Summers, M.F., Henderson, L.E., Bess, J.W. Jr., Arthur, L.O., MacDougal, J.S., Orloff, S.L., Mendeleyev, J. and Kun, E., Nature, 361 (1993) 473.
[29] De Vereese, K., Kofler-Mongold, V., Leutgeb, C., Weber, V., Vermeire, K., Schacht, S., Anne, J., De Clercq, E., Datema, R. and Werner, G., J. Virology, 70 (1996) 689.
[30] Kimura, E., Kuramoto, Y., Koike, T., Fujioka, H. and Kodama, M., J. Org. Chem., 55 (1990) 42.
[31] Kimura, E., Sakonaka, Yatsunami, T. and Kodama, M., J. Am. Chem. Soc., 103 (1981) 3041.
[32] Kimura, E., Kodama, M. and Yatsunami, T., J. Am. Chem. Soc., 104 (1982) 3182.
[33] Kataoka, M., Naganawa, R., Odashima, K., Umezawa, Y., Kimura, E. and Koike, T., Anal. Lett., 22 (1989) 1089.
[34] Fujioka, H., Koike, T., Yamada, N. and Kimura, E., Heterocycles, 42 (1996) 775.
[35] Koike, T., Takashige, M., Kimura, E., Fujioka, H. and Shiro, M., Chem. Europ. J., 2 (1996) 617.
[36] Ryser, H.J. -P., Levy, E.M., Mandel, R. and Disciullo, G.J., Proc. Natl. Acad. Sci. USA, 91 (1994) 4559.
[37] Datema, R., Rabin, L., Hincenbergs, M., Moreno, M.B., Warren, S., Linquist, V., Rosenwirth, B., Seifert, J. and McCune, J.M., Antimicrob. Agents Chemother., 40 (1996) 750.
[38] Koike, T., Takamura, M. and Kimura, E., J. Am. Chem. Soc., 116 (1994) 8443.
[39] Dessolin, J., Gala, P., Vlieghe, P., Chermann, J. and Kraus, J., J. Med. Chem., 42 (1999) 229 and references therein.

CHEMISTRY OF PLATINUM ANTICANCER DRUGS

Jorma Arpalahti

Perspectives on Bioinorganic Chemistry
Volume 4, pages 165–208.

ISBN: 0-7623-0352-2

1. INTRODUCTION

Two milestones in chemistry may be addressed to the geometric isomers of diamminedichloroplatinum(II). About 100 years ago Alfred Werner laid down the principles of modern coordination chemistry with the help of isomeric $[PtCl_2(NH_3)_2]$ [1]. About 70 years later Barnett Rosenberg and co-workers discovered that the *cis* isomer is antitumor active, whereas the *trans* isomer is not [2,3]. Today *cis*-$[PtCl_2(NH_3)_2]$ (cisplatin) is routinely used as a drug, particularly effective against testicular and ovarian cancer [4].

Soon after the discovery of the antitumor activity of cisplatin it was realized that platinum binding to nucleic acids plays an important role in the action of the drug. This led to an enormous research boom on the interactions of Pt with nucleic acids constituents. Because of the severe toxic side effects of cisplatin (nausea, nephrotoxicity, neurotoxicity, and vomiting), numerous other platinum compounds have been synthesized and tested for biological activity [4]. However, only a few of them have proven useful as drugs. One example of less toxic platinum drugs is carboplatin, where the two Cl^- ligands are replaced by the bidentate cyclobutanedicarboxylate group (Figure 1). In addition to Pt(II) compounds, efforts were concentrated also on various Pt(IV) complexes. Recently, a class of orally administered drugs containing Pt(IV) center have been developed. The compound JM–221 (*cis,trans,cis*-$[PtCl_2(NH_3)(C_6H_{11}NH_2)(OC(O)C_3H_7)_2]$ is a typical member of this class (Figure 2) [4].

Based on early structure–activity relationships the *cis* geometry was considered necessary for biological activity of Pt compounds. However, recent findings have shown that also *trans*-Pt derivatives may be active as antitumor drugs [4–6]. It appears that actually none of the classical structure–activity relationships is absolutely necessary, viz. *cis* geometry with the general formula *cis*-$PtX_2(amine)_2$ for Pt(II) and *cis*-$PtX_2Y_2(amine)_2$ for Pt(IV), anionic leaving ligand X with intermediate binding strength, and at least one NH moiety in the amine ligands [4].

Recent reviews in the field of platinum anticancer drugs focus on platinum–nucleobase chemistry [7], biological processing of platinum-modified DNA [8], *trans*-platinum anticancer drugs [5], cisplatin and derived anticancer drugs [4,9], proteins and cisplatin [10], *trans*-diammineplatinum(II) and nucleic acids [11], and catalytic activity and DNA [12], just to mention a few. The aim of this review is to explore the chemistry in the interaction of various platinum compounds with nucleic

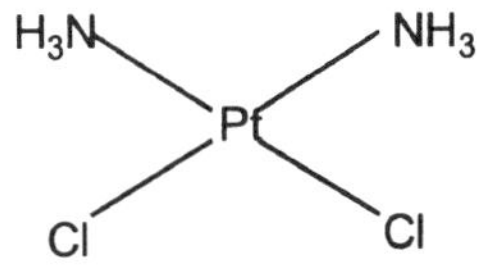

cis-$[PtCl_2(NH_3)_2]$,
cisplatin, *cis*-DDP

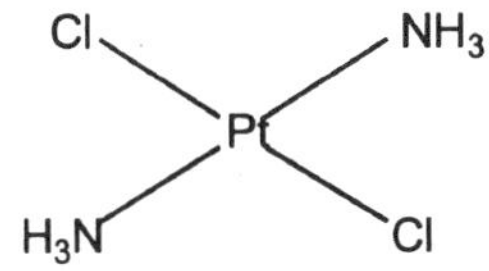

trans-$[PtCl_2(NH_3)_2]$,
transplatin, *trans*-DDP

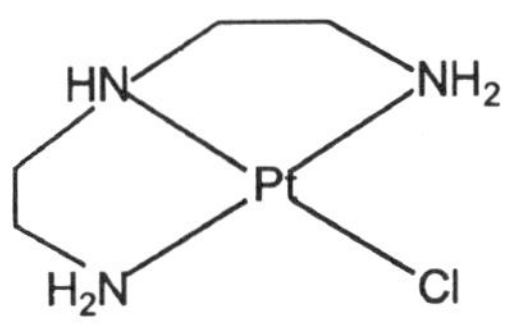

[PtCl(dien)]Cl

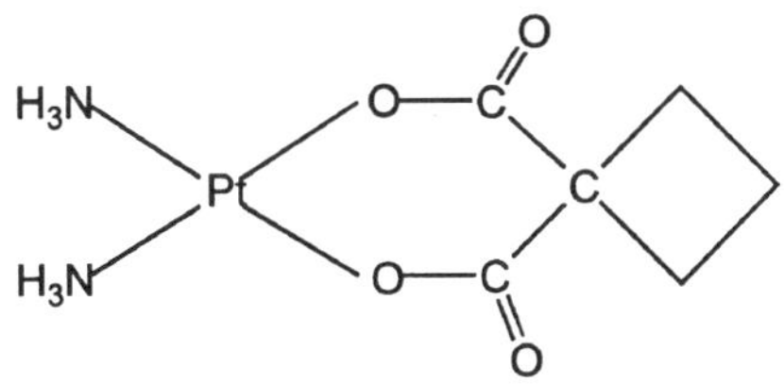

Diammine(cyclobutanedicarboxylato)platinum(II),
carboplatin

Figure 1. Schematic structures of *cis*- and *trans*-DDP, carboplatin and the model compound [PtCl(dien)]Cl.

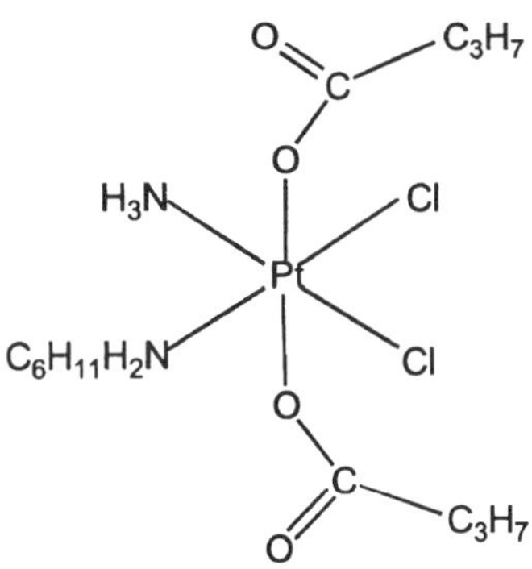

JM-221

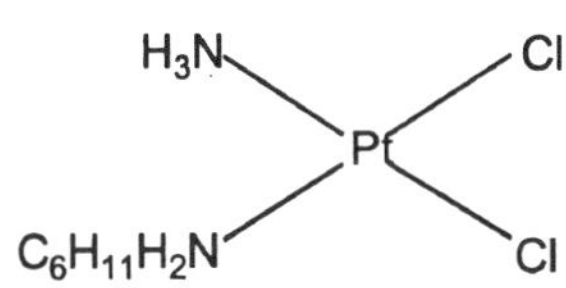

cis-$[PtCl_2NH_3(NH_2C_6H_{11})]$

Figure 2. Schematic structures of a Pt(IV) drug JM-221 and its metabolite *cis*-$[PtCl_2(NH_3)(C_6H_{11}NH_2)]$.

acids and their constituents giving special emphasis on reaction kinetics and mechanisms.

After this brief introduction, the chapter will focus on solvolysis reactions and acid–base equilibria of various platinum compounds and on species distribution of isomeric $[PtCl_2(NH_3)_2]$ in aqueous solution in Section 2. Binding of platinum compounds to monomeric nucleobase derivatives will be discussed in Section 3, while Section 4 pays attention to the reactions of Pt–nucleobase complexes with different nucleophiles. And finally, the interactions of Pt with DNA and defined oligonucleotides will be discussed in Section 5.

2. FUNDAMENTAL AQUEOUS CHEMISTRY

A common factor to cisplatin related compounds is the coordination sphere of platinum consisting of two tightly bound nitrogen-donor ligands and two leaving groups. The latter are usually relative labile Cl^- ions or oxygen donor ligands. The same seems to hold truth also for the metabolites of orally administered Pt(IV) drugs [13]. According to the existing view, these labile groups are replaced by heterocyclic nitrogen atoms of the nucleobases when the drugs bind to DNA [4,8]. On the other hand, they may also be replaced by other biomolecules resulting in a general toxicity [8]. The usual mechanism for substitution reactions of Pt(II) compounds involves a solvent path and a nucleophile-dependent path [14]. In the former, the leaving group is first displaced by the solvent molecule, after which the incoming ligand displaces the solvent molecule. In aqueous solution, in particular, the solvent path plays a major role in substitution reactions of Pt(II) with weak nucleophiles, such as nucleobases, which has important consequences regarding the active species of the platinum drugs.

2.1 Hydrolysis Reactions

In aqueous solution the labile ligands (X^-) are stepwise replaced by water molecules. Under acidic conditions the hydrolysis reactions are strongly reversible, which gives an equilibrium mixture according to Eq. 1. Much effort has been given to quantify the rate and equilibrium

$$[Pt(amine)_2X_2] \underset{k_{-1}}{\overset{k_1}{\rightleftharpoons}} [Pt(amine)_2X(H_2O)]^+ \underset{k_{-2}}{\overset{k_2}{\rightleftharpoons}} [Pt(amine)_2(H_2O)_2]^{2+} \tag{1}$$

constants for various Pt(II) compounds, particularly for *cis*- and *trans*-DDP [11,15]. According to the order of *trans*-effects $Cl^- > NH_3 > H_2O$, the first hydrolysis step is faster for *trans*-DDP, while the hydrolysis of the second Cl^- ligand is faster for *cis*-DDP. For the same reason, the Cl^- anation reactions obey the order $k_{-2}(cis) > k_{-2}(trans)$ and $k_{-1}(trans) > k_{-1}(cis)$ [15]. Even though the values reported for the individual rate parameters in various papers are quite different, data for the equilibrium constants described by Eq. 1 are in reasonable agreement $\{K_1(cis) \approx 3 \times 10^{-3}$ M, $K_1(trans) \approx 3 \times 10^{-4}$ M, $K_2(cis) \approx 4 \times 10^{-4}$ M, $K_2(trans) \approx 2 \times 10^{-5}$ M$\}$ [15]. Thus, the first hydrolysis step is more favorable for *cis*-DDP and in both cases the first step is thermodynamically more favorable than the second one.

Replacement of the NH_3 group(s) in *cis*-DDP with other amines affects only little the hydrolysis rate of the Cl^- ligand(s) (Table 1). Accordingly, the rate constant for the hydrolysis of the first Cl^- ligand from *cis*-$[PtCl_2(NH_3)(NH_2C_6H_{11})]$ (Figure 2) is comparable to that of

Table 1. Rate Constants for the Stepwise Cl^- Hydrolysis $\{k_i/(10^{-4}\ (s^{-1})\}$ and Cl^- anation $\{k_{-i}/(10^{-4}\ M^{-1}\ s^{-1})\}$ Reactions of Various Pt(II) Compounds[a]

Compound	k_1	k_{-1}	k_2	k_{-2}	*T/(K)*
cis-$[PtCl_2(NH_3)_2]$	0.25	(76)[b]	0.33	(825)	293.2
	0.632	62.6	(0.25)	909	293.2
	0.76	(195)			303.2
	0.766[c]				310
	5.13	454			318.2
	1.9	600	2.3	9800	318.2
trans-$[PtCl_2(NH_3)_2]$	0.98	(3000)	<0.5		293.2
	1.2	5000			303.2
	10.5	22000	0.04 (0.2)[d]	2000	318.2
cis-$[PtCl_2(NH_3)(NH_2C_6H_{11})]$[e]	0.19				298.2
$[PtCl_2(en)]$	0.34	(155)	0.44	3100	298.2
rac-$[PtCl_2(R_1\text{-en})]$[f]	0.32	440	0.78	6700	298.2

Notes: [a]Refer to Eq. 1. Data from [15] unless otherwise stated.
[b]The data in parenthesis refer to calculated values.
[c]From [94].
[d]At pH 11 for the hydrolysis of *trans*-$[PtCl(OH)(NH_3)_2]$.
[e]From [16].
[f]From [17].

cis-DDP [16]. Similarly, the rate parameters for the hydrolysis of a series of substituted ethylenediamine Pt(II) complexes, *rac*-$[PtCl_2(R_i\text{-en})]$ (Figure 3), can be compared to those of $[PtCl_2(en)]$ [17], and they do not markedly differ from the data given for *cis*-DDP. The same holds truth also for the Cl^- anation reactions of the aqua derivatives of these species [17]. Similar behavior might be expected also for the novel drugs exhibiting *trans* geometry of the two Cl^- ligands. However, solvolysis reactions in DMSO have shown that substitution of the NH_3 group(s) with aromatic nitrogen donor ligand(s) decreases the rate of solvolysis as compared to *trans*-DDP [5]. The effect is most profound for *trans*-$[PtCl_2(pyridine)_2]$, which shows solvolysis reaction only at forcing conditions, e.g. in the presence of Ag^+ ions. Also hydrolysis reactions of these species follow this trend, which has been attributed to steric effects of the aromatic ligand [5].

In acidic aqueous solution the displacement of 1,1-cyclobutanedicarboxylate ($cbdca^{2-}$) from $[Pt(cbdca)(NH_3)_2]$ (Figure 1) resembles the successive displacements of two monodentate carboxylates [18]. Rate constants of $8 \times 10^{-5}\ s^{-1}$ and $8.0 \times 10^{-4}\ s^{-1}$ (298.2 K) were found for the ring-opening step of nonprotonated and protonated species, respectively.

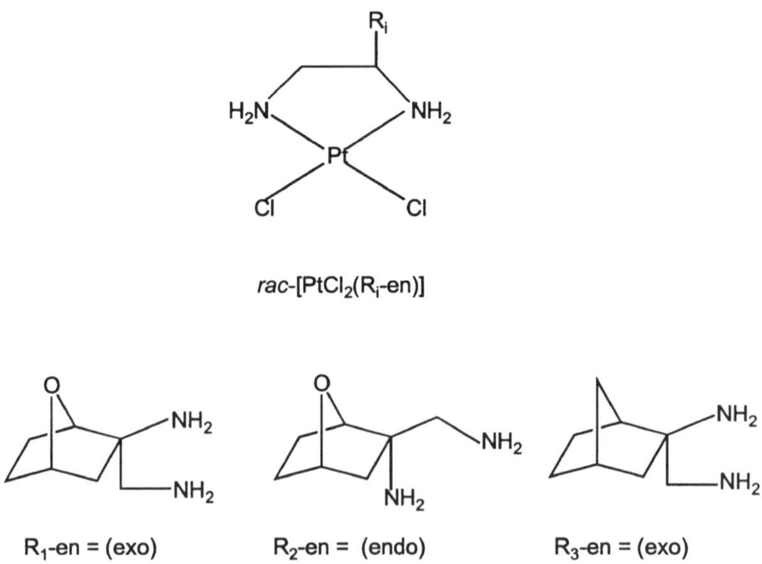

Figure 3. Schematic structures of subtituted ethylenediamine Pt(II) complexes, *rac*-$[PtCl_2(R_i\text{-en})]$.

Without added acid or nucleophile the lability of the starting compound is very low (<1 × 10^{-8} s^{-1}) most probably because of the efficient ring-closing process [18]. The second step depends linearly on the pH with a rate constant of 1.61 × 10^{-4} M^{-1} s^{-1}. In the presence of added Cl^-, the final product of the hydrolysis is *cis*-DDP although the relevance of similar process *in vivo* was questioned [18]. It has been reported that carboplatin bound to DNA retains the dicarboxylate group, probably as a monodentate ligand [19].

2.2. Acid–Base Equilibria

The hydrolysis products of $[PtCl_2(amine)_2]$ type of compounds undergo acid–base equilibria shown in Scheme 1. The pK_a values of various Pt(II) compounds are given in Table 2. Comparison of the data for aquated *cis*- and *trans*-DDP shows that the pK_a values for the *trans* isomer are about one logarithmic unit smaller in the case of diaqua and chloroaqua species. By contrast, practically similar pK_a values have been reported for the monoaquamonohydroxo species of these isomers.

The diaqua form of *cis*-DDP can be easily obtained from the dichloro compound by using Aq^+, whereas the corresponding *trans*-DDP derivative requires more forcing conditions [15]. Neither of these species have been isolated, however. Various procedures have been employed to obtain solutions of the first hydrolysis or solvolysis product of *cis*-DDP and related compounds including chromatographic methods [15] and treatment of the dichloro compound [16] with 1 equiv Aq^+ in DMF. With *trans*-DDP, controlled hydrolysis of the parent compound in basic solution gives *trans*-$[PtCl(OH)(NH_3)_2]\cdot H_2O$ and *trans*-$[Pt(OH)_2(NH_3)_2]\cdot$

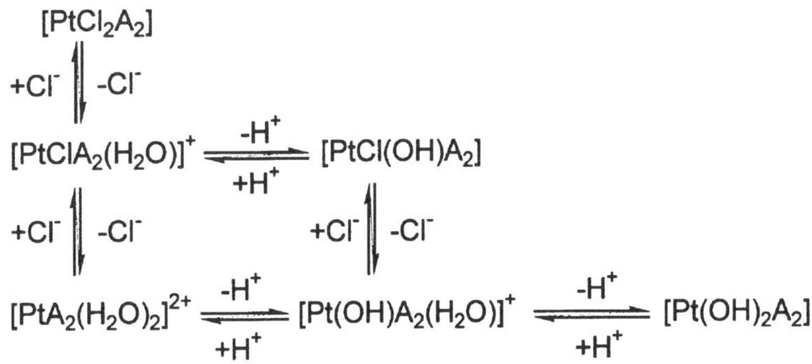

Scheme 1.

Table 2. Acid Dissociation Constants of Various Aquated Pt(II) Amines[a]

Compound	pK_a
cis-$[Pt(NH_3)_2(H_2O)_2]^{2+}$	5.37; 5.56; 5.64
cis-$[Pt(OH)(NH_3)_2(H_2O)]^{+}$	7.21; 7.32; 7.40
cis-$[PtCl(NH_3)_2(H_2O)]^{+}$	6.41; 6.85
trans-$[Pt(NH_3)_2(H_2O)_2]^{2+}$	4.32; 4.35
	4.48[b]
trans-$[Pt(OH)(NH_3)_2(H_2O)]^{+}$	7.38; 7.40
	7.20[b]
trans-$[PtCl(NH_3)_2(H_2O)]^{+}$	5.63
	5.94[c]
$[Pt(dien)(H_2O)]^{2+}$	6.13; 6.24; 6.53
	6.0[d]
$[Pt(en)(H_2O)_2]^{2+}$	5.8
$[Pt(OH)(en)(H_2O)]^{+}$	7.6
cis-$[PtCl(NH_3)(NH_2C_6H_{11})(H_2O)]^{+}$	6.4[e]

Notes: [a]T = 293–300 K. Data from [15] unless otherwise indicated.
[b]From [41].
[c]From [34].
[d]From [53].
[e]From [16].

$2H_2O$ [15,20]. The first solvolysis product of *trans*-DDP in DMSO has also been isolated and characterized with different counterions [11].

The hydrolysis products of *cis*- and *trans*-DDP are known to undergo condensation reactions with formation of di- and oligo-μ-OH species [11,15]. However, for geometric reasons less robust OH bridged species are expected for *trans* compounds compared to the cyclic species formed by the *cis* isomer [11].

2.3. Distribution Diagrams

The equilibrium constants given above for the hydrolysis and protolysis reactions of *cis*- and *trans*-DDP can be employed to construct distribution diagrams of various species as a function of the pH. In human blood plasma ($[Cl^-]_{ambient} \approx 0.1$ M) the dichloro species predominates at about pH 7.4 for both *cis*- and *trans*-DDP (Figure 4). By contrast, in intracellular conditions ($[Cl^-]_{ambient} \approx 0.004$ M) the hydrolysis products dominate, but the distribution behavior of the two isomers is quite

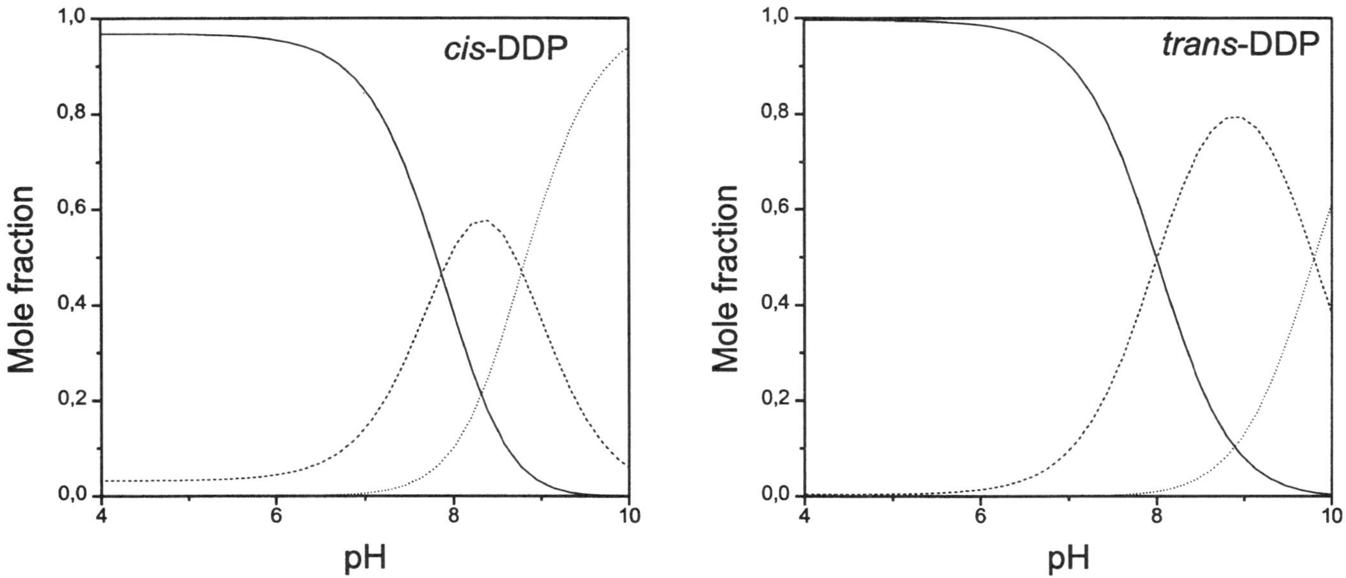

Figure 4. Distribution diagrams for *cis*- and *trans*-DDP as a function of the pH in extracellular conditions ($[Cl^-]_{ambient} \approx 0.1$ M). Notation as X, Y in $[Pt(NH_3)_2X,Y]$: solid line (Cl^-), dashed line (Cl^-, H_2O), dotted line (H_2O, H_2O).

different (Figure 5). With *cis*-DDP, the mole fraction of various $[Pt(NH_3)_2X,Y]$ species at pH 7.4 are Cl^-,Cl^- (0.04), Cl^-,H_2O (0.03), Cl^-,OH^- (0.31), OH^-,H_2O (0.31), OH^-,OH^- (0.31). In the case of *trans*-DDP, the corresponding data are Cl^-,Cl^- (0.12), Cl^-,H_2O (0.01), Cl^-,OH^- (0.72), OH^-,H_2O (0.08), OH^-,OH^- (0.07). In both cases the mole fraction of $[Pt(NH_3)_2(H_2O)_2]^{2+}$ form is negligible. At pH 7.4, the predominant species for *cis*-DDP is the second hydrolysis product, which constitutes >60% from the total amount of Pt and is present as the monoaquamono-hydroxo and dihydroxo forms in a ratio of about 1:1. In the case of *trans*-DDP, instead, the first hydrolysis product dominates over the diaqua species. At pH 7.4 the former is almost completely in the chloro-hydroxo form, while the latter is present as the monoaquamonohydroxo and dihydroxo forms in a ratio of about 1:1. The distribution diagrams for *cis*-DDP correspond to those presented earlier [21]. The minor differences in the relative amounts of *cis*-$[PtCl(NH_3)_2(OH)]$ and *cis*-$[Pt(NH_3)_2(OH)(H_2O)]^+$ at pH 7.4 (Figure 5) may be accounted for the different $[Cl^-]_{ambient}$ values (3.5 mM in [21]) employed in the calculations. However, for a number reasons these findings should be applied with care in assigning the actually active species of *cis*- and *trans*-DDP inside the cell, as pointed out earlier [7].

3. PLATINUM-NUCLEOBASE INTERACTIONS

3.1 Binding Sites

The availability of different metal ion binding sites in 9-substituted purine and pyrimidine nucleobases and their model compounds has been recently reviewed by Lippert [7]. The distribution of metal ions between various donor atoms depends on the basicity of the donor atom, steric factors, interligand interactions, and on the nature of the metal. Under appropriate reaction conditions most of the heteroatoms in purine and pyrimidine moieties are capable of binding Pt(II) or Pt(IV) [7]. In addition, platinum binding also to the carbon atoms (e.g. to C5 in 1,3-dimethyluracil) has been established [22]. However, the strong preference of platinum coordination to the N7 and N1 sites in purine bases and to the N3 site in pyrimidine bases cannot completely be explained by the negative molecular electrostatic potential associated with these sites [23]. Other factors, such as kinetics of various binding modes and steric factors, appear to play an important role in the complexation reactions of platinum compounds.

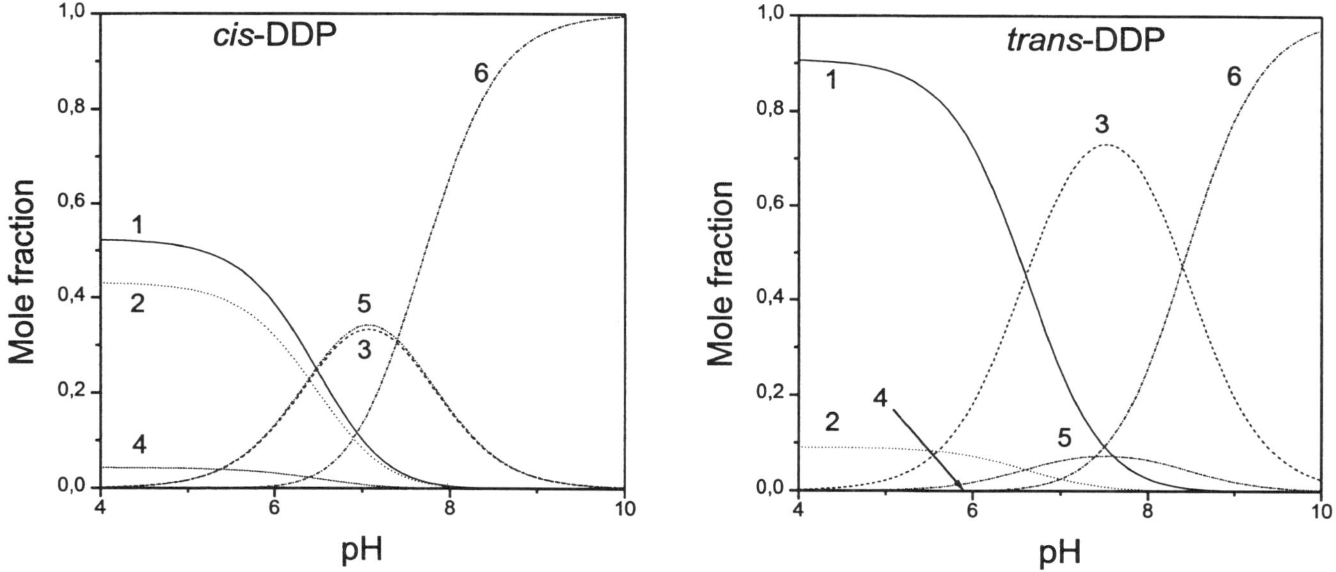

Figure 5. Distribution diagrams for *cis*- and *trans*-DDP as a function of the pH in intracellular conditions ($[Cl^-]_{ambient} \approx 0.004$ M). Numbering of the species as X, Y in $[Pt(NH_3)_2X,Y]$: 1 (Cl^-, Cl^-), 2 (Cl^-, H_2O), 3 (Cl^-, OH^-), 4 (H_2O, H_2O), 5 (OH^-, H_2O), 6 (OH^-, OH^-).

Purine Derivatives

The predominant binding site in 9-substituted 6-oxopurines (guanine and hypoxanthine derivatives) is the N7 atom of the base (Figure 6). The prevailing keto tautomer requires proton at N1 even in mildly acidic conditions, which efficiently prevents platination of the N1 site [7]. Under neutral and basic conditions competition of Pt(II) between the N1 and N7 sites has been reported. Attachment of Pt(II) to the N7 atom acidifies the N1H proton and facilitates coordination of additional platinum ions to both N1 and N3 [7]. In N7,N9-blocked 6-oxopurines, the N1 site is the major coordination site [7,24].

Comparing to 6-oxopurines, 9-substituted adenine derivatives exhibit a more versatile binding behavior. Distribution of various Pt(II) compounds between the adenine N1 and N7 sites has been intensively studied [7,15]. At low pH, N7 coordination predominates due to protonation of the N1 site ($pK_a \approx 3.8$ [25]). Above pH 4, various platinum compounds either slightly favor the N7 site over the N1 site or distribute almost equally between these sites depending on the adenine derivative studied [15]. In Pt excess, 9-substituted adenines easily form also N1,N7-diplatinated complexes [7]. The slight preference of the N7 site suggests that steric and kinetic factors associated with this binding mode are able to compete with more favorable electronic effects of the N1 site. Nevertheless, the Pt–N1 bond appears to be thermodynamically stronger than the Pt–N7 bond. Equilibration of the mixture of aquated Pt^{II}(dien) and adenosine at elevated temperature has been shown to significantly increase the amount of the N1 bound complex at the expense of the N7 isomer [26]. Coordination of platinum to other sites in 9-substituted adenines is rare. The N3 binding mode has been found for Pt^{II}(dien)

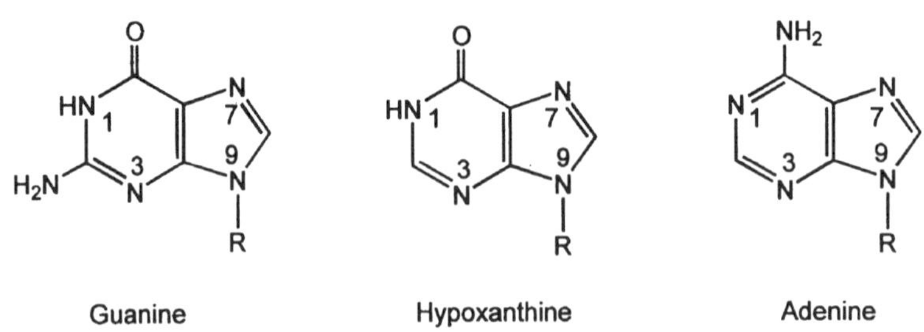

Figure 6. Schematic structures of common purine nucleobases with the numbering scheme.

binding to N6′,N6′,N9-trimethyladenine, i.e. when both N1 and N7 are sterically blocked [27]. Simultaneous binding of Pt(II) to N1 and N6 with loss of a proton has been proposed in the reaction of 4-picoline(2,2′:6′,2″-terpyridine)platinum(II), a potential intercalator of poly[d(A-T)$_2$] rather than anticarcinogenic drug, with adenosine and 2′-deoxyadenosine [28]. After initial N1 platination, a loss of a proton from the C(6)–NH_2 group leads to subsequent rapid platination of N6, which is further facilitated by stacking of the terpyridine moieties. Exceptionally, Pt(II) can form a chelate involving N1 and deprotonated C(6)–NH_2 group, as observed in the reactions of *cis*-$Pt^{II}(PMe_3)_2$ with 9-alkylated adenines under neutral conditions [29,30]. In acidic solution, the N1 bound biscomplex is formed [30]. The higher thermodynamic stability of the Pt–N1 bond over the Pt–N7 bond may account for, at least partly, the unexpected preference of the N1 site in *cis*-$Pt^{II}(PMe_3)_2$–adenine complexes by taking the strong *trans*-effect of phosphorous into account.

Pyrimidine Derivatives

The coordination properties of pyrimidine bases seem to be less versatile than those of purine derivatives. Various Pt(II) and Pt(IV) compounds, including *cis*- and *trans*-DDP, preferentially bind to the N3 site in N1-substituted cytosine derivatives (Figure 7), as verified by a variety of methods [7]. Simultaneous binding to N3 and to the exocyclic amino group C(4)–NH_2 upon loss of a proton has been observed in a bridged Pt(II) system and in a chelated Pt(IV) system [7]. With 1,3-dimethyluracil, Pt(II) coordination to the C5 atom has been ascertained by X-ray crystallography [22].

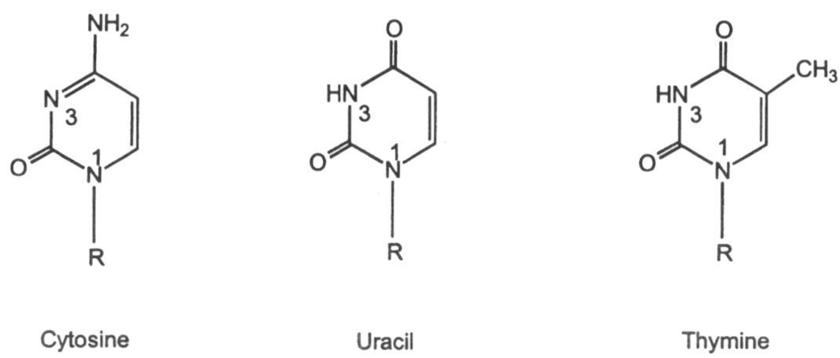

Figure 7. Schematic structures of common pyrimidine nucleobases with the numbering scheme.

Platination of the N3 position in 1-substituted uracil and thymine derivatives requires proton abstraction and usually occurs only at high pH, but the Pt–N3 bond, once formed, is thermodynamically stable (log $K \approx 9.6$) [7]. Platinum binding to N3 increases the basicity of O4, which becomes an additional binding site leading to di- and trinuclear complexes. A list of X-ray structurally characterized species is given by Lippert [7]. Pt complexes of uracil and thymine can form intensely colored adducts (e.g. platinum pyrimidine blues), which show anticarcinogenic activity analogously to the monomeric species [7].

3.2 Kinetic Studies

Direct substitution of the chloro ligand(s) with nucleobases is a relative slow process in *cis*-DDP, *trans*-DDP, and related compounds [31], and usually the rate-determining step in the complexation is the Cl^- hydrolysis reactions. For example, *cis*-DDP shows almost no selectivity between adenosine and inosine with second-order rate constants of 1.6×10^{-3} $M^{-1}s^{-1}$ (313.2 K) and 1.9×10^{-3} $M^{-1}s^{-1}$ (318.2 K), [15] respectively. Similarly, rate constants reported for *cis*-DDP binding to adenosine and guanosine nucleotides are very similar ranging from 6.25×10^{-3} $M^{-1}s^{-1}$ to 9.33×10^{-3} $M^{-1}s^{-1}$, though this data should be considered with caution for the reasons given elsewhere [15]. The reactivity of *trans*-DDP towards inosine is higher than that of *cis*-DDP (relative rate constant about 5 at 318.2 K), in line with the *trans*-effect $Cl^- > NH_3$ [31]. By contrast, the corresponding 1:1 complexes {*cis*- and *trans*-$[PtCl(NH_3)_2(Ino\text{-}N7)]^+$} have an equal tendency to bind the second inosine [31]. A rapid direct substitution (within minutes) of the Cl^- ligand with various nucleotides has been found in the reaction of the ring-opened species of *cis*-$[Pt\{Me_2N(CH_2)_2PPh_2\text{-}N,P\}_2]Cl_2$ [32] (a cytotoxic Pt complex, Figure 8) under physiological conditions and this compound exhibits similar affinities for guanine and thymine moieties [33]. In addition to the strong *trans* effect of phosphorous, the removal of the N(3)H proton from the thymine moiety enhances the platination reaction. It is further noted that this Pt(II) compound is able to bind to the thymine base in double-stranded d(TTGGCCAA) [33].

In acidic aqueous solution the complexation of the chloroaqua derivatives of *cis*- and *trans*-DDP involves substitution of the aqua ligand with the nucleobase [15]. The *trans* derivative reacts 10 times faster than the *cis* isomer with inosine at pH 3, in line with the *trans*-effect $Cl^- > NH_3$. It has been shown that *trans*-$[PtCl(NH_3)_2(H_2O)]^+$ behaves like a mono-

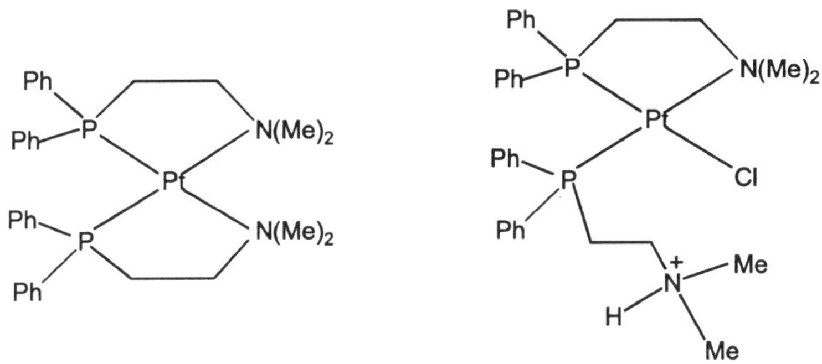

I II

Figure 8. Schematic structures of a cytotoxic Pt complex *cis*-$[Pt\{Me_2N(CH_2)_2PPh_2\text{-}N,P\}_2]Cl_2$ (**I**) and its ring-opened species (**II**).

functional platinum(II) species in the pH range 2.8–8.4 (298.2 K) [34]. Throughout this pH range the complex formation with inosine derivatives has been explained by substitution of the aqua ligand with the incoming nucleoside, the OH appearing to be inert toward substitution. Although no pH-dependent data appear to exist for the corresponding *cis* derivative, the Cl^- hydrolysis step may compete with complexation via aqua ligand under neutral and slightly basic conditions, at least partly. In the *cis* isomer, the lower reactivity of the aqua ligand combined with faster hydrolysis step relative to the *trans* isomer may partially cancel the larger fraction of the reactive *cis*-$PtCl(NH_3)_2(H_2O)]^+$ species. Although the chloroaqua derivative of *cis*-DDP seem to favor 6-oxopurines over other nucleobases, this effect is more profound in the complexation of the diaqua species [15].

With aquated Pt(II) compounds, numerous studies have revealed the kinetic preference of the 6-oxopurine N7 site [15,35]. In addition to the favorable electrostatic potential mentioned above [23] also steric factors seem to favor coordination to the guanine N7 site, in particular [36]. Estimated relative steric parameters (in parenthesis) suggest that the guanine N7 (1.00) and hypoxanthine N7 (1.03) atoms are the least sterically hindered binding sites in alkylated nucleobases, followed by the adenine N7 (1.17) and deprotonated hypoxanthine N1 (1.17) sites and the deprotonated N3 atoms of the different pyrimidine bases (1.39 for U, 1.44 for T, and 1.56 for C), while the adenine N1 (1.58) and deprotonated guanine N1 (1.61) sites are the most hindered ones. Rate parameters for stepwise complexation of *cis*-$[Pt(NH_3)_2)(H_2O)_2]^{2+}$ with various nucleobase derivatives clearly reveal the kinetic preference of Pt(II) to 6-oxopurine derivatives [15]. It has been suggested that the substituent at C6 of the purine moiety plays an active role in the complexation. The oxo substituent enhances the reaction rate by forming an H-bond to the aqua or am(m)ine group bound to Pt(II). By contrast, the amino group at C6 sterically hinders the attack of Pt(II) to both N1 and N7, in agreement with steric predictions presented above. In the guanine moiety, the amino group at C2 prevents platination of the (deprotonated) N1 site [15].

Much attention has been paid to the enhanced reactivity of purine-5′-mononucleotides, in particular, which has been attributed to the electrostatic interactions between the phosphate group and platinum center, or to hydrogen bonding between phosphate and the amine or aqua ligand bound to platinum [15,35]. However, no such intramolecular interactions have been confirmed in solid state with mononucleotides until recently [37]. In $[Pt(en)(5'\text{-GMP-N7})_2]\cdot 9H_2O$, macrochelate rings are formed via intramolecular H-bonding between monoanionic 5′-phosphate group and coordinated ethylenediamine NH (Figure 9). This macrochelation is present also in solution both at pH 7 (dianionic nucleotide) and at pH 2–3 (monoanionic nucleotide) [37]. Interestingly, the X-ray structurally characterized complex $[Pt(en)(5'\text{-GMP})_2]\cdot 3H_2O$ contains only intermolecular H-bonding [38]. In the nonahydrate as well as in the corresponding Pd(II) complex, eletrostatically bonded axial water molecules play key roles in a network of H-bonding involving the phosphate oxygens, ethylenediamine NH, and the C(6)O group of the nucleotide [37]. Similar hydrogen bonding system involving Pt–NH_2 group, 5′-phosphate oxygen and the C(6)O group of the base is present also in the complex $[Pt([^{15}N_3]dien)(5'\text{-GMP-N7})]^{2+}$, as suggested from two-dimensional $^{1}H,^{15}N$ NMR studies [39].

Far less data are available from other diaqua Pt(II) compounds. Comparison of the diaqua derivatives of *cis*- and *trans*-DDP has shown that their complexation with inosine derivatives is mechanistically similar, but the rate parameters for various steps show considerable differences [40,41]. For example, for isomeric $[Pt(NH_3)_2(H_2O)_2]^{2+}$ ions $k_1(cis) \approx 10\, k_1(trans)$, whereas for the $[Pt(OH)(NH_3)_2(H_2O)]^+$ ions the difference is $k_1(cis) \approx 6\, k_1(trans)$ in the formation of 1:1 complexes. The ability of isomeric 1:1 complexes to bind the second nucleobase is, however, very similar in both cases, also by taking proton transfer formally from inosine

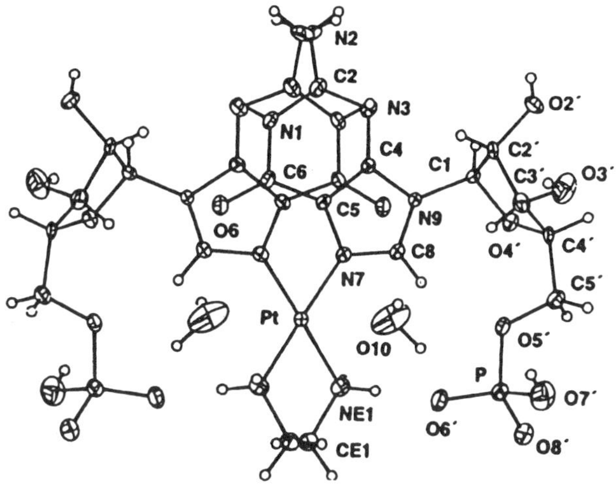

Figure 9. Molecular structure of [Pt(en)(5′-GMP-*N*7)$_2$·9H_2O showing the macrochelate ring formation via intramolecular H-bonding. (Reproduced with permission from ref. 37).

N(1)H to the deprotonated OH bound to Pt(II) into account. This proton transfer seems to play a significant role in enhancing the complex formation at high pH [41].

Although the kinetics of Pt(IV) compounds are generally much slower than those of Pt(II), significant rate enhancement for Pt(IV) may be achieved in the presence of catalytic amounts of the corresponding Pt(II) derivative. For example, in the reaction of 9-methylhypoxanthine with *cis*-$[PtCl_4(NH_3)_2]$ a 30-fold rate enhancement was observed upon addition of 10% of *cis*-$[PtCl_2(NH_3)_2]$ [42]. The "chloride bridging" mechanism [14] probably accounts for the catalytic effect of Pt(II) observed for other Pt(II)/Pt(IV) pairs, too [42]. In addition, a catalytic effect of Pt(II) has been reported for the reaction of iproplatin (CHIP) with ascorbic acid at pH 7.0 (iproplatin = *cis*-dichloro-*trans*-dihydroxo-*cis*-bis(isopropylamine)platinum(IV) [43].

3.3 Effects of Platinum Binding

Coordination of an electrophilic platinum to the ring atoms of the nucleobases withdraws electron density from the ring. As a result, the heteroatoms of neutral nucleobases capable of deprotonation become more acidic and those capable to accept a proton become less basic. For example, the N(1)H proton of 9-substituteted 6-oxopurines is acidified by 1.2–2.0 log units upon platination of the N7 site depending predominantly on the charge of the platinum compounds [7,34,40]. With 9-substituted adenines, the pK_a of the N1 site and that of the exocyclic NH_2 group is lowered about 2 and 4 log units due to N7 platination, respectively [7]. Even more dramatic change has been observed for N1,N7-diplatinated 9-methyladenine, where the pK_a of the NH_2 group is lowered about 6 log units [7]. On the other hand, an increase in basicity is observed when Pt displaces a proton upon coordination to nucleobases. In the case of inosine, $[Pt(dien)]^{2+}$ binding to N1 makes the N7 site about 1.1 log units more basic [44]. With 1-substituted uracil and thymine derivatives, an increase of 4–5 log units in the pK_a of the exocyclic oxygens results from Pt binding to the N3 site [7].

The binding of platinum to the endocyclic nitrogens of purine nucleosides may change the hydrolytic stability of the N-glycosidic bond, particularly with 2′-deoxyribonucleosides [45]. Binding of $[Pt(dien)]^{2+}$ to the N7 site of dIno and dGuo enhances spontaneous cleavage of the N-glycosidic bond, but retards acid-catalyzed depurination. In the case of N7 platinated dAdo, hydrolysis is retarded only at $pH < 2$. Instead, N1 platination of these species does not significantly affect their acid-catalyzed depurination. Most drastic effect is observed for N1,N7-diplatinated dIno and dAdo, where the depurination is markedly accelerated in the presence of $[Pd(dien)]^{2+}$. It has been proposed that the catalytic effect of Pd results from the coordination of $[Pd(dien)]^{2+}$ to the N3 site of the nucleobases [45].

Hydrogen bonding between complementary bases G and C as well as A and T plays a fundamental role in double-stranded DNA. The attachment of Pt to a nucleobase may change the hydrogen-bonding pattern for any of the following reasons: blocking of H-bonding sites by the metal, template distortion, DNA cross-linking, *anti-syn* switch, pK_a shift, tautomer equilibrium shift, or generation of rare tautomers [46]. A very recent review focuses on the effects of metal ion binding to nucleobase pairing [47]. Usually, base pairing between G and C is according to a Watson-Crick scheme. Very rarely pairing is of the reverse Watson-Crick type or of Hoogsteen type [48]. If the N1 and N7 sites of guanine are blocked, hydrogen bonding with cytosine is via N^2 and N^3 as shown in Figure 10 [48].

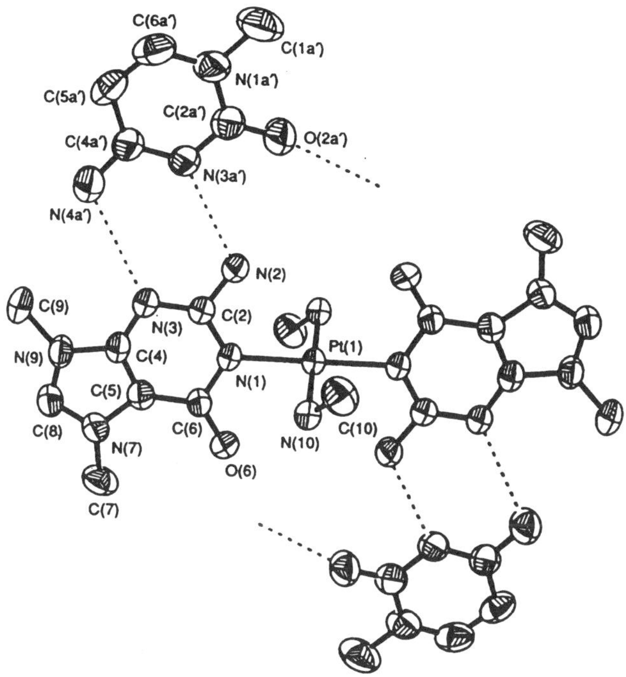

Figure 10. Hydrogen bonding scheme between 7,9-dimethylguanine (7,9-dmgua) and 1-methycytosine (mcyt) in *trans*-$[Pt(MeNH_2)_2(7,9\text{-dmgua-N1})_2(ClO_4)_2 \cdot 2mcyt$ with two hydrogen bonds, N(2)···N(3a′) 2.881(8) Å and N(3)···N(4a′) 3.194(8) Å. (Reproduced with permission from ref. 48).

4. REACTIONS WITH SULFUR LIGANDS

Various sulfur-containing (bio)molecules play important roles in biological processing of platinum anticarcinogenic drugs. For example, thiols or thioethers such as sodium thiosulfate or sodium diethylditiocarbamate reduce the nephtrotoxic side effects of *cis*-DDP and small peptides like glutathione or metallothionine may prevent Pt binding to DNA [4]. In addition, sulfur-containing molecules are used as trapping agents in studying platinum binding to nucleic acids fragments [8]. The high affinity of Pt(II) for a sulfur atom probably accounts for the usefulness of these compounds. Although the Pt–S bond is relatively inert and thermodynamically stable, the sulfur donor can be displaced from Pt coordination sphere even with nitrogen ligands which may have important implications in the active mechanism of Pt drugs.

Reactions of the anticancer drug carboplatin with sulfur-containing amino acids have shown that thiols react very slowly forming sulfur-bridged species containing four-membered Pt_2S_2 rings as the predominant products [49]. In contrast, reactions with thioethers are more rapid. Surprisingly, L-methionine forms very stable ring-opened species which has a half-life of 28 h for Met-N,S ring closure at 310 K [49].

The presence of L-methionine increases the rate of reaction of *cis*-DDP with 5′-GMP [50]. The major reaction pathway proposed involves initial attack of methionine to *cis*-DDP through the sulfur atom. After N,S-chelation of L-methionine, the nucleotide displaces the NH_3 group at pH 7 owing to the *trans* labilization effect of the sulfur atom. At the initial stage of the reaction displacement of monodentate sulfur-bound L-methionine with 5′-GMP was also observed. The displacement of S-bound methionine may be even more favorable in biological systems. Isolation of a Pt–bis(methionine) complex from the urine of patients treated with *cis*-DDP suggests potentially important role of L-methionine in the metabolism of Pt anticancer drugs [50]. Similarly, 5′-GMP is able to remove S-bound L-methionine from the complex $[Pt(dien)(Met\text{-}S)]^{2+}$ [4, 51]. The rate constants for reaction of 5′-GMP with $[Pt(dien)(Met\text{-}S)]^{2+}$ ($5.1 \times 10^{-5}\ M^{-1}s^{-1}$, pH 7.0, $T = 298$ K [51]) is close to the apparent rate constants for its reaction with $[Pt(dien)Cl]^{+}$ ($6.2 \times 10^{-5}\ s^{-1}$, pH 5.0, $T = 295$ K [52], in which the rate-limiting step is the hydrolysis of the chloro ligand. Interestingly, the reaction of $[Pt(dien)Cl]^{+}$ with 5′-GMP appears to be mechanistically unclear. Recent ^{15}N NMR studies have shown no peaks for the aqua derivative in aqueous solution of $[Pt(dien)Cl]^{+}$ after 7 days at 298 K [53]. It has been suggested that the reaction of this Pt(II) species with GMP does not involve hydrolysis, or the hydrolysis is catalyzed by the nucleotide [53].

Intramolecular replacement of sulfur by nitrogen has been reported in the complex of $[Pt(dien)]^{2+}$ with S-guanosyl-L-homocysteine (sgh) according to reaction 2,

$$Pt(dien)(sgh\text{-}S)]^{2+} \rightarrow [Pt(dien)(sgh\text{-}N7)]^{2+} \tag{2}$$

which nicely demonstrates the difference between kinetically and thermodynamically favored binding sites [54]. Compared to the initial formation of S-bound species ($t_{1/2} \approx 2$ h) the isomerization step is slow ($t_{1/2} \approx 10$ h). Similar slow intramolecular S $\rightarrow$ N migration of Pt^{II}(dien) ($t_{1/2} \approx 40$ h, $T = 313$ K, pH 6.5) has been observed in histidylmethionine through a dinuclear intermediate [55]. With the nucleopeptide Met-d(TpG), Pt^{II}(dien) initially coordinated to the sulfur atom of the methionine moiety migrates to the G–N7 site; the reaction was complete in 6 days at room temperature [56]. By contrast, the bifunctional Pt^{II}(en) forms a stable S,N7 chelate with Met-d(TpG) even in the presence of extra unplatinated N7 [56].

In general, the displacement of a N-bound nucleobase from the Pt coordination sphere is difficult, but can be facilitated by the attack of strong nucleophiles, e.g. CN^- and sulfur-containing ligands. However, not all platinum bound to DNA can be removed with CN^- treatment [7]. Studies with model compounds have shown that the Pt–N3 bond in thymine or uracil complexes is particularly inert toward the attack of a CN^- ion, most probably because of the protective effect of exocyclic oxo groups [7]. Substitution studies have shown that a single thymine or uracil base can protect all three other ligands in Pt(II) coordination sphere [7], except the aqua ligand [57]. The Pt–N3 bond in uracilato complexes is resistant to the attack of also other nucleophiles, e.g. thiourea [58] or I^- [57]. However, protonation of the exocyclic O4 atom of 1-methyluracil considerably increases the lability of the bis(1-methyluracilato) complex [57]. It has been suggested that involvement of the coordination sites that normally are in the interior of the duplex DNA (T–N3, C–N3, G–N1) can lead to inert cross-links [59]. Also orientation of bases affects the substitution reaction of bis(nucleobase) complexes. With Pt(II)-9-ethylguanine complexes, the head-to-tail bis(complex) resists substitution with CN^- more efficiently than the head-to-head species [59].

In excess of the nucleophile stepwise dissociation is expected for Pt(II)–bis(nucleobase) complexes according to Scheme 2, where charges are omitted for clarity. Most probably, the dissociation of the nucleobases

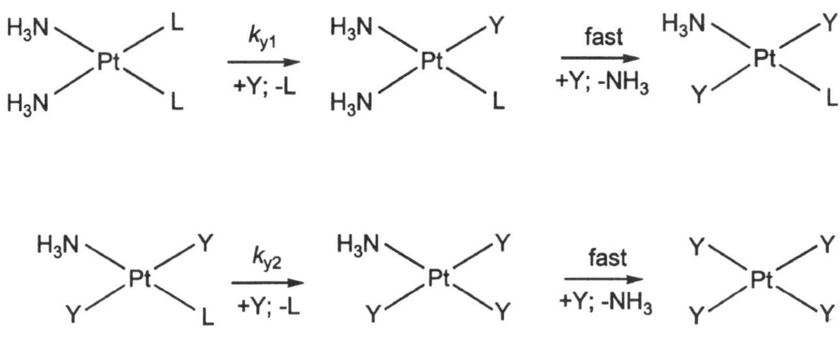

Scheme 2.

are the rate-limiting steps in the overall reaction, while the substitution of the NH_3 group *trans* to the nucleophile are fast due to the strong *trans* effect of the nucleophiles. Rate constants for the substitution reactions of different Pt(II) complexes of guanosine and adenosine in the presence of various nucleophiles (CN^-, thiourea, and I^-) are summarized in Table 3. The data indicate that steric effects have a substantial influence to the attack of the nucleophile even in closely related compounds.

Table 3. Rate Constants $\{k_i/(10^{-5}\ M^{-1}\ s^{-1})\}$ for the Thiourea Assisted Dissociation of Various Pt(II)–Nucleobase Complexes in Aqueous Solution[a]

Complex	k_{Y1}	k_{Y2}	*Exp. cond [ref.]*
cis-$[Pt(NH_3)_2(G\text{-}N7)_2]^{2+}$	5.9 ± 0.6		[b][60]
	7.96 ± 0.05	3.70 ± 0.05	[c][61]
	25 ± 4[d]		[e][62]
cis-$[Pt(NH_3)_2(G\text{-}N7)(A\text{-}N7)]^{2+}$	4.7 ± 0.1[f]	10 ± 1[f]	[g][61]
	10 ± 0.1[h]	3.3 ± 0.1[h]	
cis-$[Pt(NH_3)_2(G\text{-}N7)(A\text{-}N1)]^{2+}$	2.7 ± 0.2[f]	7.8 ± 0.1[f]	[g][61]
	3.5 ± 0.2[h]	4.1 ± 0.1[h]	
$[Pt(dien)(G\text{-}N7)]^{2+}$	82.5 ± 0.7		[i][63]
	48.0 ± 0.3[j]		[i][63]
	175 ± 1[k]	49.9 ± 0.7[l]	[m][63]
$[Pt(dien)(A\text{-}N7)]^{2+}$	24.5 ± 0.2[h]		[i][63]
	31.8 ± 0.4[j]		[i][63]
	111 ± 2[k]	88 ± 2[l]	[m][63]
cis-$[Pt(NH_3)_2(1\text{-}MeU)(H_2O)]^{+}$	37000 ± 2000		[n][57]
	6010 ± 40[j]		
cis-$[Pt(NH_3)_2(1\text{-}MeU)_2]$	no reaction[j]		[m][57]

Notes: [a]For the notation of the rate constants, see Scheme 2.
[b]$T = 316$ K, pH ≈ 4.45, I not specified.
[c]$T = 318.2$ K, pH ≈ 4.0, $I = 0.1$ M ($NaClO_4$).
[d]CN^- as the nucleophile.
[e]$T = 303$ K, pH ≈ 10, I not specified.
[f]For the dissociation of adenosine.
[g]$T = 318.2$ K, pH ≈ 4.0, $I = 0.1$ M ($NaClO_4$).
[h]For the dissociation of guanosine.
[i]$T = 318.2$ K, pH ≈ 6.5, $I = 0.1$ M ($NaClO_4$).
[j]I^- ion as the nucleophile.
[k]For the disappearance of the starting material.
[l]For the disappearance of $[Pt(dienH)(L\text{-}N7)(tu)]^{3+}$.
[m]$T = 318.2$ K, pH ≈ 3, $I = 0.1$ M ($NaClO_4$).
[n]$T = 298.2$ K, pH = 3.0, $I = 0.1$ M ($NaClO_4$).
[m]$T = 333.2$ K, pH range 4–7, $I = 0.1$ M ($NaClO_4$).

Interestingly, the rate constant for the dissociation of $[Pt(dien)(L{-}N7)]^{2+}$ (L = ado, guo) in the presence of thiourea (tu) significantly increases on going from neutral to slightly acidic aqueous solution [63]. At the same time, the mechanism of the overall dissociation is changed. In neutral solution dissociation of both complexes gives only free nucleoside and $[Pt(dien)(tu)]^{2+}$, whereas at about pH 3 the end-products are free nucleoside and $[Pt(tu)_4]^{2+}$. In acidic solutions HPLC analysis revealed the formation of an additional product in both cases, which was assigned to ring-opened species $[Pt(dienH)(L{-}N7)(tu)]^{3+}$. NMR data for isolated compounds are consistent with 4-coordinate Pt(II) species, in which the dien ligand acts as bidentate group and one of the NH_2 groups has been trapped by a proton. Although these ring-opened species are quite stable in acidic solution, they decompose back to the starting material and free ligand in a ratio of about 10:1 when the pH is increased. The rate constant for the backward reaction linearly depends on the pH; already at pH 5.5 the reaction is fast ($t_{1/2} < 3$ min at 338.2 K) [63]. The facile displacement of sulfur-bound thiourea from Pt(II) by nitrogen donor is important, as it demonstrates the nucleophilic power of a group being spatially in a favorable position.

It has been suggested that the dissociation mechanism of $[Pt(dien)(L{-}N7)]^{2+}$ in acidic solution involves pseudorotation of the 5-coordinate intermediate as shown in Figure 11, rather than spontaneous dissociation of either of the dien–NH_2 groups followed by the attachment of thiourea [63]. Activation parameters ($\Delta H^{\ddagger} = 66 \pm 5$ kJ mol^{-1}, $\Delta S^{\ddagger} = -100 \pm 15$ J K^{-1}mol^{-1} for the G–N7 complex and $\Delta H^{\ddagger} = 62 \pm 8$ kJ mol^{-1}, $\Delta S^{\ddagger} = -108 \pm 25$ J K^{-1}mol^{-1} for the A–N7 complex) are consistent with an associative substitution reaction. In addition, the observation that the I^- ion does not

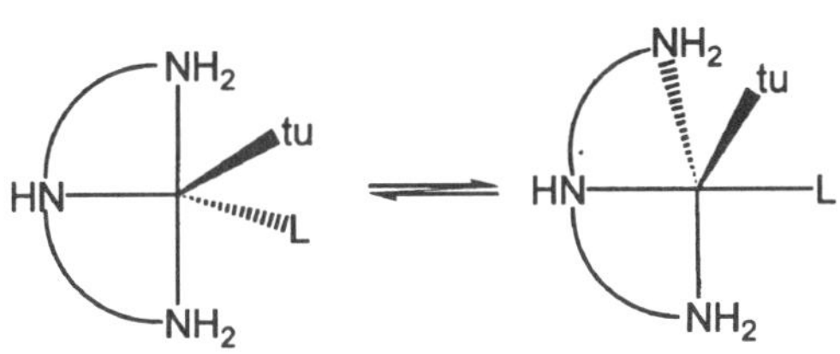

IM1 IM2

Figure 11. Proposed pseudorotation between the 5-coordinate intermediates IM1 and IM2 in the dissociation of $[Pt(dien)(L{-}N7)]^{2+}$ (L = ado, guo) in the presence of thiourea (tu).

result in ring opening of the dien ligand even at pH 1 strongly argues against the dissociation mechanism. This type of ring-opened Pt^{II}dien species with halide ions are known to be very stable at low pH, as verified by X-ray crystal structure analysis [64].

5. PLATINUM BINDING TO DNA AND DEFINED OLIGONUCLEOTIDES

5.1 Adduct Formation with *cis*- and *trans*-DDP

It is now widely accepted that DNA is the most important target for cisplatin, the biological activity of which results from its ability to bind DNA and block replication [8,9]. Several lines of evidence suggest that GG and AG intrastrand cross-links are the key adducts in the biological activity of *cis*-DDP [8]. For example, it has been shown that in *Escherichia coli*, the GG adducts are cytotoxic, while the AG cross-link is more mutagenic than the GG cross-link [8,65,66]. No *cis*-$[Pt(NH_3)_2]^{2+}$ intrastrand crosslink-specific mutations were observed for the G*TG* adduct [66]. In various studies the GG adduct has been found to account about 65% and the AG adduct 25% of the total platinum bound to DNA [4,10]. The preference for GG intrastrand cross-link is much more than the statistically expected value (37%) [4]. In order to explain the sequence selectivity of *cis*-DDP binding, numerous studies have carried by employing different oligonucleotides as model compounds.

The structural data for various *cis*- and *trans*-DDP adducts of different oligonucleotides have been compiled in recent reviews [4,8]. In GG and AG adducts (single-stranded chains) with *cis*-DDP, both nucleobases are coordinated through the N7 site to platinum in a head-to-head fashion. This orientation of bases is retained also in intrastrand GG cross-link with *cis*-DDP. The NH_3 group(s) bound to platinum may form hydrogen bonds to the C(6)O group and phosphate oxygens in intrastrand GG cross-link [4,8]. Although the attachment of platinum destabilizes the double helix, the Watson-Crick base pairs are still observed. NMR studies suggest that *cis*-DDP binding causes a bend or a kink in the double helix of 40°–70°, whereas electroforetic studies indicate a bend angle of 35°–40° [4,8].

The X-ray crystal structure of platinated duplex DNA dodecamer d(CCTCTG*G*TCTCC)·d(GGAGACCAGAGG), where G*G* represents the binding sites (G–N7) of *cis*-$[Pt(NH_3)_2]^{2+}$, is depicted in Figure 12 [67]. The coordination of Pt(II) to adjacent guanine bases distorts the

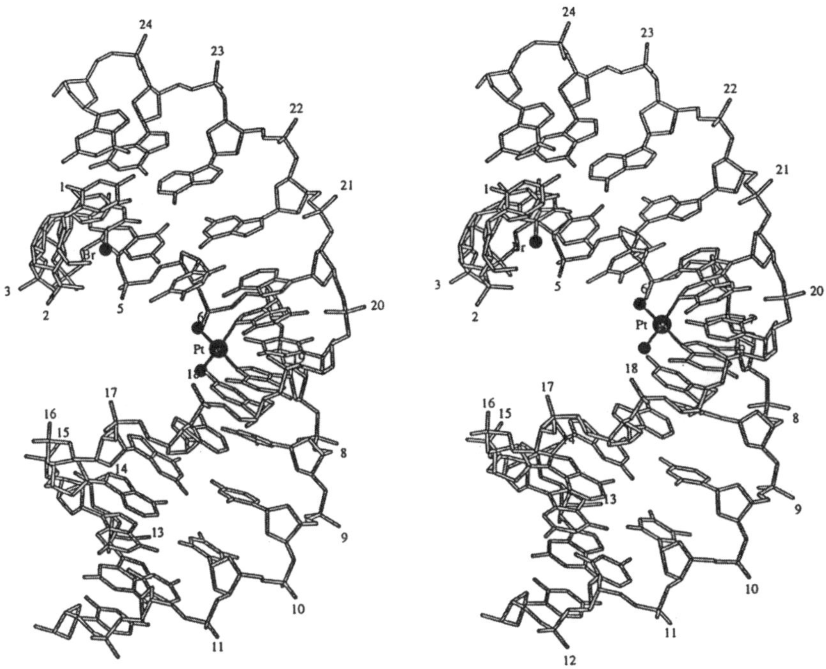

Figure 12. Stereoview of a ball and stick model of the duplex DNA, d(CCU^{Br}CTG*G*TCTCC)·d(GGAGACCAGAGG), where –G*G*– is modified by *cis*-$[Pt(NH_3)_2]^{2+}$. (Reproduced with permission from ref. 67).

double helix by causing a bend (35°–40°) toward the major groove, which is distributed over several base pairs [67]. The Pt atom lies about 1 Å out of the guanine plane (1.3 Å from the 5′-G plane and 0.8 Å from the 3-G plane) [68], because of rather small angle between the planes of the two chelating guanines. It has been estimated that the bending decreases the Pt–N7 binding energy by ca. 50 and 25 kJ/mol, respectively [68]. The structure also indicates that the DNA conformation at the 5′-side from the coordinated platinum has changed from B-DNA to A-DNA. The structural alterations caused by platinum 1,2-intrastrand cross-links (widening of minor groove and the bending of DNA) resemble those induced by cellular proteins containing the high mobility group (HMG) domain [10,67] and might explain how HMG proteins recognize cisplatin–DNA adducts [67]. The NMR solution structure of a platinated double-stranded oligonucleotide d(CCTG*G*TCC)·d(GGACCAGG) indicates a bend angle of 58° [69]. In addition, the probable hydrogen bond between NH_3 and phosphate oxygen in the crystal structure was not observed in solution structure [67]. However, both structures also show substantial similarities, e.g. similar dihedral angle between guanine bases and wide minor groove opposite to the platinum binding site [67, 69].

The solution structure of the second major *cis*-DDP adduct d(AG) has been elucidated by NMR and molecular modeling in a double-stranded nonanucleotide d(CTCA*G*CCTC)·d(GAGGCTGAG) [70,71]. Also in this case the oligonucleotide is kinked at the platinated site towards the major groove in a similar manner to that observed for the d(GG) cross-link. The major structural difference between these two adducts seems to be located in the unplatinated strand across the lesion. In the AG adduct, the complementary thymine remains stacked on the 5′-adjacent cytosine, while in the GG cross-link the base opposite the 5′-G oscillates between two positions, as shown schematically in Figure 13 [70,71]. With model nucleobases 9-methyladenine and 9-ethylguanine the two purine bases adopt a head-to-head orientation as shown by X-ray crystal structure analysis [72,73]. Depending on the counterion the orientation of the bases seems to correspond the cisplatin–AG adduct in single-stranded DNA (with NO_3^- [72]) or in double stranded DNA (with PF_6^- [73]).

In addition to intrastrand cross-link described above, *cis*-DDP can also form an interstrand cross-link representing less than 10% of the total lesions [8,74]. The NMR solution structure of the duplex d(CCTCG*CTCTC)·d(GAGAG*CGAGG) containing a single inter-

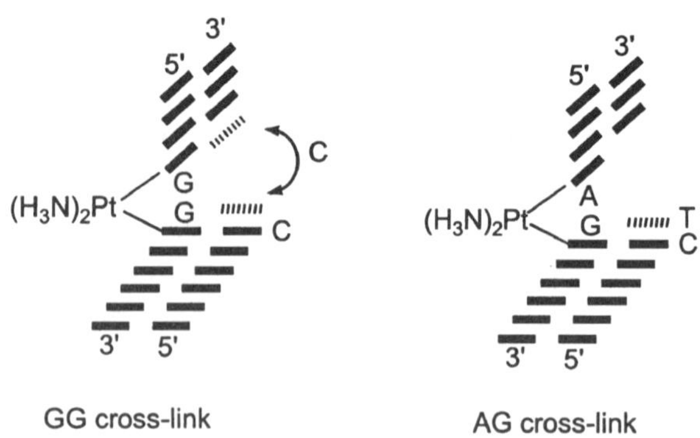

Figure 13. Schematic view showing the principal structural differences between the GG and AG coss-links [71].

strand cross-link shows that the head-to-tail arrangement of the two cross-linked guanines places the *cis*-$Pt^{II}(NH_3)_2$ unit in the minor groove [75]. The interstrand cross-link induces a 40° bend toward the minor groove. NMR data indicates an unwinding of 76°, in agreement with the values deduced from gel electrophoretic measurements [75]. With negatively supercoiled DNA at low levels of platination (<1 Pt atom fixed per 500 base pair) enhanced interstrand cross-link formation has been reported, preferentially between guanine residues in both the 5′-d(GC)-3′ and the 5′-d(CG)-3′ sites [76]. It has been suggested that superhelical turns favor contacts with topologically remote sites [68]. The interstrand cross-link at the d(G*C/G*C) site has been shown to rearrange via monofunctional adducts into intrastrand cross-links [74]. With a 10 base pair double-stranded oligonucleotide, the cleavage reaction is irreversible ($t_{1/2} \approx 29$ h) and the complementary strands separate after the cleavage of the interstrand cross-link. Within a longer cross-linked oligonucleotide (20 base pairs), the interstrand cross-link is more stable ($t_{1/2} \approx 120$ h), probably because the interstrand cross-link cleavage is reversible [74]. It has been further noted that the intrastrand cross-link at the d(GG) site and especially the interstrand cross-link at the d(GC/GC) site are photosensitive, which results in a cleavage of Pt–G bonds [77].

Both *cis*-DDP and [Pt(dien)Cl]Cl are able to form monofunctional adducts that produce local denaturation in DNA. The denaturational change preferentially occurs in the base pair containing platinated guanine in a sequence d(Py-G-Py), where py is a pyrimidine base [78]. Monoadduct formation at the sequence d(5′-GC-3′), at which interstrand cross-links are formed, also causes denaturational change in DNA [79]. Since denaturation of DNA has not been observed upon platination at the major sequences d(GG) and d(AG), conformational alterations may facilitate the formation of minor bifunctional Pt–DNA adducts [78]. At low ionic strength ($[Na^+] = 1$ mM), instead, various Pt(II) compounds {*cis*-DDP, *trans*-DDP and [Pt(dien)Cl]Cl} increase the melting temperature of DNA, but this effect is reversed at higher Na^+ concentration [80]. With *cis*-DDP-modified DNA, however, the reversal occurs at markedly lower $[Na^+]$ values than for other Pt(II) compounds studied. Based on these studies the thermal stability of Pt-modified DNA is attributed to stabilization effects of the positive charge on Pt and of interstrand cross-links, and a destabilization effect of conformational distortions in DNA [80].

Compared to *cis*-DDP far less information is available for the clinically ineffective *trans* isomer. Early studies have shown that *trans*-DDP binds to DNA and forms bifunctional adducts [4,8,12]. Most favorable binding targets of *trans*-DDP in DNA include G–N7 and A–N7 in 1,3-intrastrand adducts, and C–N3 in 1,4-intrastand adducts, whereas 1,2-intrastrand adducts are not possible for geometric reasons. The 1,3-intrastand adducts of *trans*-DDP through the N7 sites of adenine and guanine are capable of duplex formation, although with base pair disruptions near the platination site in a double-helical structure [4,8]. With double-stranded DNA the bisadduct spectrum of dG-Pt-dC (50%), dG-Pt-dG (40%), and dG-Pt-dA (10%), quite different from that of *cis*-DDP, has been reported [8]. However, at low level of platination, *trans*-$[Pt(NH_3)_2\{d(G^*XG^*\}]$ cross-links have not been detected in native DNA; Pt binding mainly gave monofunctional adducts and interstrand cross-links [81]. A molecular modeling study has been carried out with a *trans*-DDP adduct within double-stranded oligonucleotide d(TCTG*TG*TC)·d(GACACAGA). At this stage, however, it has not been possible to discriminate between the three families of intrastrand adducts proposed [82]. Further difference between *cis*- and *trans*-DDP is that the latter forms interstrand cross-links preferentially between complementary G and C residues [12]. Even though the duplex is distorted on both sides of the interstrand cross-link, the bases are still paired [12].

Although the bifunctional lesions of *cis*- and *trans*-DDP are considered stable under physiological conditions, a few exceptions are known. The 1,3-intrastrand cross-link in *trans*-$[Pt(NH_3)_2\{d(CG^*CG^*)\}]$ rearranges into the 1,4-*trans*-$[Pt(NH_3)_2\{d(C^*GCG^*)\}]$ intrastrand cross-link [8]. Similar rearrangement has also been reported for sequences d(CGAG) and d(CGTG); the rate of rearrangement being independent of pH in the range of 5–9 ($t_{1/2}$ = 120 h at 303 K) [12]. It appears that intervening base in the d(CGXG) sequence has no major effect on the reaction, whereas the C residue on the 5′ side plays a key role in the rearrangement reaction. However, other studies have not found this type of rearrangement in d(CGAG) sequence [8,12]. In oligonucleotides containing *trans*-$[Pt(NH_3)_2\{d(G^*XG^*)\}]$ intrastrand adducts (X = A, T, C), a rearrangement into interstrand cross-links appears as soon as the platinated oligonucleotides are hybridized with their complementary ribonucleotide or deoxyribonucleotide strands ($t_{1/2} \approx 6$ h at 310 K) [12]. Rearrangement of Pt–G adducts has been reported also for *cis* geometry. The *cis*-$[PtNH_3)_2\{\text{-}G^*G^*\text{-}\}]$ intrastrand bisadduct in the duplex d(CCTG*G*TCC)·d(GGACCAGG) has been found to isomerize at room temperature in NaCl solution into the interstrand bisadduct d(CCTGG*TCC)·d(G*GACCAGG) [69]. It has been suggested that this isomerization proceeds via chloro-monoadducts since it does not occur in Cl^- free medium [68,69].

Rearrangement of platinum 1,3-intrastrand cross-link into interstrand cross-links may be a new tool for selective modulation of gene expression. It has been shown that *trans*-DDP forms 1,3-*trans*-$[Pt(NH_3)_2)(GNG)]$ cross-links with single-stranded oligo(2′-O-methylribonucleotide)s, which are inert within the single-stranded chain [83,84]. When paired with their complementary RNA strands the platinum intrastrand cross-links rearrange into interstrand cross-links, and the rate of this reaction appears to depend upon the sequence facing the intrastrand cross-links. The conversion into interstrand cross-link may be achieved in a few minutes when the triplets 5′-CN′C (N′= any nucleotide, $t_{1/2} \geq 3$ h at 310 K) are replaced by a doublet 5′-UA or 5′-CA [83]. Also the relative position of platinum affects the rearrangement; it does not occur when the 1,3-intrastrand cross-link is near the end of the double helix [84].

A single transplatin monofunctional adduct (*trans*-$[PtCl(NH_3)_2(dC)]^+$ or *trans*-$[PtCl(NH_3)_2(dG)]^+$) within a homopyrimidine oligonucleotide can form an interstrand cross-link when the platinated oligonucleotide is bound to the complementary duplex [85]. In all triplexes formed, the interstrand cross-link between a platinated Hoogsteen strand and the complementary purine-rich duplex predominantly involved G residues. Instead, interstrand cross-links were not detected with noncomplementary DNA [85]. The fastest cross-linking reaction was observed with *trans*-$[PtCl(NH_3)_2(dC)]^+$ adduct containing Hoogsteen strand ($t \approx 2$ h), whereas the adduct *trans*-$[PtCl(NH_3)_2(dG)]^+$ at the 5′ end of the Hoogsteen strand gave the slowest cross-linking reaction ($t \approx 6$ h). It was pointed out that direct platination of a single G or C residues in an oligonucleotide containing several C or G bases is difficult [85]. Recent achievements on automatic solid-phase synthesis, however, facilitate the preparation of site-specifically platinated oligonucleotides, as shown recently for the purine-rich sequence d(TCAGGTAG*G*ACTTGGTGTCT) bearing a *cis*-$[Pt(NH_3)_2]^{2+}$ unit [86] and for *trans*-DDP-treated sequences d(CXCA) and d(ATAGTAXACAGA), where X denotes the *trans*-$[Pt^{II}(NH_3)_2(T\text{–}N3)]$ unit [87].

5.2. Adduct Formation with Other Platinum Compounds

The compound *cis*-[$PtCl_2(NH_3)(C_6H_{11}NH_2)$] (Figure 2) forms with DNA an intrastand cross-link involving adjacent guanosine residues as the major adduct (54%) [13]. Other major adducts detected include interstrand or long-range intrastrand cross-links between guanine moieties (18%) and d(AG) intrastrand adducts (8%). The different amines in a Pt coordination sphere give rise to two orientational isomers of platinated d(GG) moiety, both of which were detected in platination of DNA and shorter oligonucleotides [13].

Recently, several groups have reported on antitumor activity of platinum compounds exhibiting *trans* geometry [5,6,88,89]. Bifunctional compounds *trans*-[$PtCl_2$(E-iminoether)$_2$] {iminoether = HN=C(OMe)–Me} preferentially form monofunctional adducts at guanine residues in double-helical DNA [90,91]. These monofunctional adducts convert into bifunctional adducts much more slowly than in the case of *trans*-DDP, indicating that the reactivity of the second leaving group is markedly reduced. The isomeric [$PtCl_2$(E-iminoether)$_2$] can produce the B $\rightarrow$ Z transition in DNA [92]. The *cis* isomer (*cis*-EE) affects the B $\rightarrow$ Z transition in a similar way than *cis*-DDP, whereas the *trans* isomer (*trans*-EE) affects the B $\rightarrow$ Z transition only slightly. The behavior of *trans*-EE seems to be, however, fundamentally different from that of *trans*-DDP, which hinders the B $\rightarrow$ Z transition. These findings seem to be consistent with a unique binding mode of this antitumor *trans* compound [92].

Binding studies of dinuclear platinum compounds, [{*trans*-($PtCl(NH_3)_2$}$_2${μ-$H_2N(CH_2)_nNH_2$}]$^{2+}$ (1,1/t,t) (Figure 14) to oligonucleotides containing a central d[TG*G*T) site have shown that the (Pt,Pt)-intrastrand adducts result in a flexible nondirectional bend, different from that induced by *cis*-DDP [93]. In addition, this bend is essentially independent of the length of the diamine linker in these dinuclear Pt species (n = 2,4,6). While the cisplatin adduct displays a 100% N-type sugar of the 5′-G and an *anti* base conformation of the platinated bases, the dinuclear adduct does not display the typical N-type sugar pucker [93]. Rather, the base orientations are *anti* (5′-T), *anti* (G1), *anti*/*syn* (G2), and *anti* (3′-T) and the sugar conformations are N, S/N, N, and S, respectively.

With the dinucleotide d(GG), initial binding of the dinuclear Pt(II) compounds (1,1/t,t; n = 2–6) is relatively fast compared to the ring closure into the macrochelate adduct [94]. In addition, the rate of ring

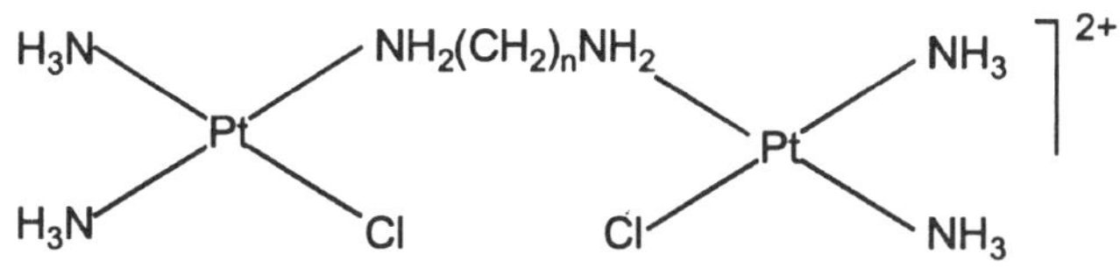

1,1/c,c

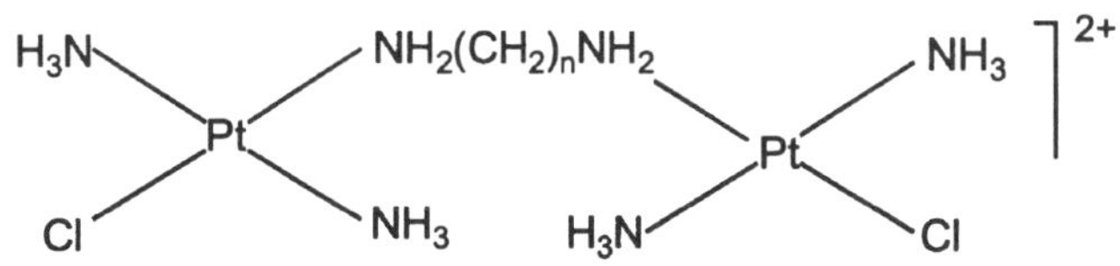

1,1/t,t

Figure 14. Schematic structures of isomeric dinuclear platinum compounds, $[\{(PtCl(NH_3)_2\}_2\{\mu\text{-}H_2N(CH_2)_nNH_2\}]^{2+}$ (n = 2–6).

closure depends on the length of the amine linker being faster when n = 4–6. The presence of a *syn*-orientated G base in macrochelates (n = 3 or 6) is different from the *cis*-DDP chelate and may explain the flexible bending of these dinuclear Pt(II) species in DNA [94]. Comparison of the reactivity of Pt(II) dinuclear species of the type (1,1/t,t; n = 4,6) and (1,1/c,c; n = 4,6) (Figure 14) has shown that the initial binding to r(GG) was slower with 1,1/c,c than with 1,1/t,t $\{t_{1/2} \approx 1.2$ h (n = 4) and 1.0 h (n = 6) for 1,1/c,c, and $t_{1/2} \approx 0.8$ h (n = 4, 6) for 1,1/t,t at 310 K$\}$ [95]. Instead, the closure into bisadduct was faster for the 1,1/c,c species $\{t_{1/2} \approx 6.5$ h (n = 4) and 4.2 h (n = 6) for 1,1/c,c, and $t_{1/2} \approx 11.0$ h (n = 4) and 6.4 h (n = 6) for 1,1/t,t at 310 K$\}$. In the d(GG)-1,1/c,c (n = 4) adduct the 3′-G has a *syn* base orientation and 60% S-type/40% N-type sugar conformation, while the 5′-G has an *anti* base orientation and S-type sugar conformation [95]. Reactions with natural DNA in a cell-free medium have shown that the diplatinum species form with a relatively short half-time (≈ 1 h) an interstrand cross-link as the major adduct [96]. At higher levels of modification also intrastrand cross-link formation between adjacent bases may be possible.

The ligand substitution reactions of Pt(IV) are slow compared to those of Pt(II), which makes their direct interaction with DNA less likely. Rather, the antitumor activity of Pt(IV) compounds has been suggested to require *in vivo* reduction to the more reactive Pt(II) derivatives [97, 98]. However, it has been reported that the compound oxoplatin {*cis*-diamminedichloro-*trans*-dihydroxoplatinum(IV)}, a Pt(IV) analogue of *cis*-DDP, binds directly to DNA without addition of a reducing agent [99]. Although the binding of oxoplatin is slow compared to *cis*-DDP, the binding sites in DNA appear to similar in both cases [99].

An example of an antitumor active Pt compound lacking the NH group is $[PtCl_2(bmic)_2]$ {bmic = bis-(N-methylimidazol-2-yl)carbinol), Figure 15} [100]. This compound binds to G residues in a similar fashion than *cis*-DDP, except that it may form stereoisomers as demonstrated by its reaction with d(GG). The preference of certain stereoisomer as well as the presence of many conformers in solution has been attributed to H-bonding from the OH group to the C(6)O groups of the guanine bases.

A completely different mode of interaction has been reported for $Pt(en)_2$, which binds to double-stranded oligonucleotides in a noncovalent manner. NMR studies have indicated that in the duplexes $d(TCGGGATCCCGA)_2$ and $d(CAATCCGGATTG)_2$ this square planar Pt(II) species selectively binds to the AT sequence in the minor groove [101]. It was suggested that the binding is stabilized by close van der Waals interactions as well as hydrogen bonds [101].

5.3. Kinetic Studies

Although structural studies on various platinated oligonucleotides clearly demonstrate the preponderance of GG and AG adducts, they do

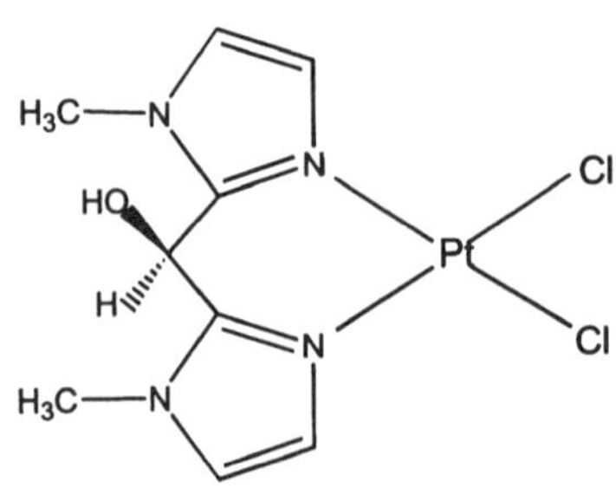

Figure 15. Schematic structure of $[PtCl_2(bmic)_2]$, an antitumor active Pt compound lacking the NH group.

not give an unambiguous answer for the reasons of this preference. Like in simple model compounds, kinetics of different binding modes is often cited to cause the predominant formation of GG and AG adducts. Early studies already indicated that the relative rates of formation of different adducts are invariant with time [8]. This, together with the general inertness of Pt–N bonds, makes the initial platination step very crucial because coordination of cisplatin and related compounds to DNA is a two-step process. The first step involves the formation of monofunctional adducts, primarily at the N7 position of guanine or adenine. These monoadducts then react further to form intrastrand and interstrand cross-links [4,8,12]. Much attention has been paid to the formation of Pt–DNA adducts with *cis*- and *trans*-DDP, in particular, to explain the differences in their biological activity.

Provided that the hydrolysis of the first Cl^- ligand is the rate-determining step in the reaction of *cis*- and *trans*-DDP with DNA, the formation of monofunctional DNA adduct with *trans*-DDP should be faster (Section 2.1). However, the rate parameters for the first platination step of DNA are very similar for both isomers ($t_{1/2} \approx 2$ h, pH 6.5 or 7.4) [8,12]. This apparent discrepancy may be attributed to the side-reactions of *trans*-$[PtCl(NH_3)_2(H_2O)]^+$ (viz. Cl^- anation and deprotonation of the aqua ligand), which more efficiently decrease the concentration of the reactive monoaqua species of *trans*-DDP as compared to the *cis* isomer [4]. Based on data from model studies a slightly longer lifetime of Pt–DNA monoadducts is expected for *trans*-DDP than for *cis*-DDP. Indeed, hydrolysis of the second Cl^- ligand seems to be the rate-limiting step in the closure of monofunctional adducts, which is only slightly faster for the cis isomer [8,12]. With *trans*-DDP, also considerably longer lifetimes for monoadducts have been reported [8,12,102]. The observed differences in lifetimes may result from experimental problems in trapping the monoadducts [8]. In addition, it has been shown that closure of *trans*-DDP monoadducts is strongly affected by the presence of other adducts and by the length of the DNA fragment [102]. It has been concluded that within *trans*-DDP-modified DNA, at low level of platination, the monofunctional adducts evolve slowly ($t_{1/2} > 24$ h) into bifunctional lesions and that these bifunctional lesions are mainly interstrand cross-links. This has been suggested to explain, at least partly, the clinical inefficiency of *trans*-DDP [102].

The experimental determination of the rate constants for the platination of different oligonucleotides has turned out to be rather difficult. To avoid the complicacy due to the Cl^- hydrolysis of the dichloro species, kinetic studies are usually carried out with aquated Pt derivatives. Since these are difficult to detect during the course of the reaction, various trapping methods are used to monitor different reaction steps [8,103]. In one type of approach, HPLC trapping method has been used to study the reactions of the aquated *cis*-DDP with oligonucleotides [103,104,105]. The employment of $^{1}H,^{15}N$ NMR spectroscopy offers another way to study platination reactions of oligonucleotides. By employing ^{15}N labeled *cis*-DDP the major species of platination of both single- and double-stranded GG oligonucleotides can all be detected simultaneously by $^{1}H,^{15}N$ NMR spectroscopy [106]. This method appears to be very sensitive and allows the direct detection of the aquachloro intermediate present at only micromolar concentrations.

The second major problem with aquated platinum derivatives deals with the reaction conditions (pH and ionic strength). In slightly acidic solution, where the kinetic measurement usually are carried out, an increase in the pH decreases the concentration of the reactive aquachloro and diaqua species (Section 2.2), thereby decreasing the rate constant. This can be avoided by employing suitable biological buffers incapable of metal ion coordination (e.g. sulfonic acid derivatives) [41,107]. By contrast, the employment of phosphate buffers in large excess should be avoided [8]. In the case of oligonucleotides, in particular, the ionic strength of the mixture may play a significant role in the complexation with Pt(II) because of condensation of cations present in the solution on the surface of polyanion. It has been shown that rate constant for the binding of *cis*-$[PtCl(NH_3)(NH_2C_6H_{11})(H_2O)]^+$ to the sulfur atom of phosphorothioate containing oligonucleotides $d(T_4p(S)T_4)$ and $d(T_8p(S)T_8)$ significantly decreases when the Na^+ concentration of the medium increases [16,108]. However, similar effect was not observed for the corresponding dinucleotide d(Tp(S)T). It is noteworthy that the aquated platinum derivative, as a positive ion, may also associate on the surface of the oligonucleotide. Thus, the formation of the final adduct may consist of two pathways. The first route goes via direct attack of Pt to the final target and the second one involves reversible preassociation steps(s) and directed diffusion of Pt along the chain into the final binding site [16].

In general, covalent coordination of platinum to phosphate groups has been neglected at the oligonucleotide level, although such interaction is feasible and has been observed with phosphate anions [109]. In *N,N*-dimethylformamide, *cis*-DDP seems to be able to coordinate to the phosphodiester groups of $d(TpT)^-$ and $d(TpG)^-$ [110]. In the case of $d(TpT)^-$, quantitative coordination of Pt to the nonbridging phosphoryl oxygen atoms was observed. With d(TpG)$^-$, instead, formation of a macrochelate was proposed in which the guanine N7 atom and either of the terminal phosphoryl oxygens are coordinated to platinum [110].

Rate constants $k_{5'}$ and $k_{3'}$ (Scheme 3) for the first complexation step of various oligonucleotides with *cis*-$[Pt(NH_3)_2(H_2O)_2]^{2+}$ and related compounds are summarized in Table 4. Broadly speaking, the rate constants listed in Table 4 for various oligonucleotides do not markedly differ form the values given for various mononucleotides [15]. For the reasons outlined above, direct comparison of data in Table 4 is, however, difficult because of different experimental conditions and methods employed. Nevertheless, certain general comparisons can be made. The rate constants reported for various ribonucleotides XpG and GpX (X = G, A, C) with *cis*-$[Pt(NH_3)_2(H_2O)_2]^{2+}$ indicate a slight preference of the 3′-guanine over the 5′-guanine at ambient ionic strength [111]. In addition, a guanine at the place X accelerates the first platination step. For $[Pt(NH_3)_3(H_2O)]^{2+}$, the preference for G is much weaker. This behavior has been attributed to hydrogen bonding from the aqua ligand to the C(6)O group, which favors the first coordination of the diaqua derivative [15,111]. In the reactions of GG-containing single-stranded oligonucleotides with *cis*-$[Pt(NH_3)_2(H_2O)_2]^{2+}$ or $[Pt(NH_3)_3(H_2O)]^{2+}$, similar rate

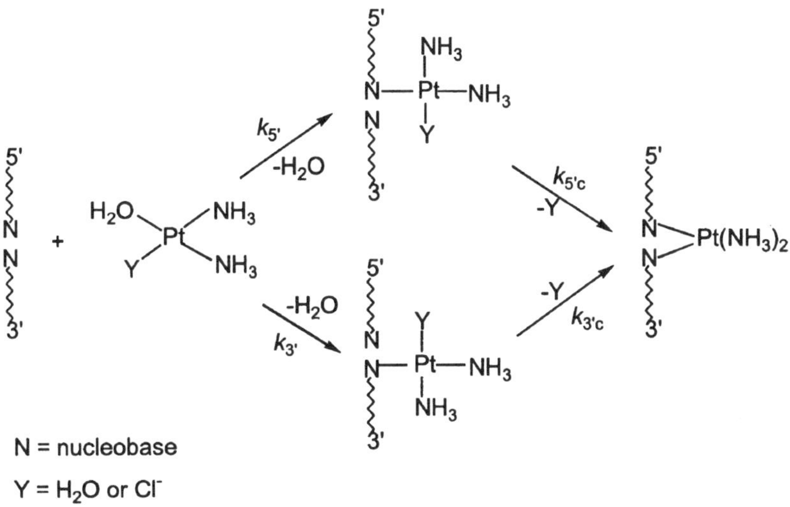

Scheme 3.

Table 4. Second-Order Rate Constants for Binding of *cis*-$[Pt(NH_3)_2(H_2O)_2]^{2+}$ to the G–N7 Sites of Various Oligonucleotides

Oligonucleotide[a]	$k_{5'}/M^{-1}\ s^{-1}$	$k_{3'}/M^{-1}\ s^{-1}$	*Exp. cond. [ref.]*
GG	2.8	5.7	[b] [111]
AG		1.7	[b] [111, 112]
		0.171	[c] [113]
CG		2.2	[b] [111]
GA	0.8		[b] [111, 112]
GC	0.95		[b] [111]
d(GG)	0.39	0.41	[d] [104]
d(TGG)	0.87	0.87	[d] [104]
	0.90	0.89	[e] [103]
d(GTG)	[f]	21.6	[g] [114]
d(TTGG)	2.1	1.7	[d] [104]
d(CTGGCTCA)	4.2	2.0	[d] [104]
d(ATACATGGTACATA)ss[h]	4.5[i], 1.43[j]	4.8[i], 2.3[j]	[k] [105]
d(ACATGGTACA)ss[h,l]	0.84[m]		[n] [106]
d(GG)[o]	0.026[m]		[p] [16]
$d(T_7GGT_7)$[p]	0.92[m]		[p] [16]

Notes: [a]Direction of the chain 5′→3′, see Scheme 2. Charges are omitted for clarity.
[b]T = 293.2 K, pH 5.2, I not specified.
[c]T = 291.5 K, pH 3.3, I not specified.
[d]T = 293.2 K, pH 4.4, I = 0.1 M ($NaClO_4$).
[e]T = 293 K, pH 4.4.
[f]$k_{5'} << k_{3'}$.
[g]T = 310.2 K, pH 6.27–6.43, I not specified.
[h]Single-stranded chain.
[i]From NMR data.
[j]From HPLC data.
[k]T = 288 K, pH 4.8, I = 0.1 M ($NaClO_4$).
[l]Data for *cis*-$[PtCl(NH_3)_2(H_2O)]^+$.
[m]Platination side not defined.
[n]T = 310 K, pH 7.1.
[o]Data for *cis*-$[PtCl(NH_3)(NH_2C_6H_{11})(H_2O)]^+$.
[p]T = 298.2 K, pH 6.50.

constants were found for the platination of 5′- and 3′-guanines in d(GG) and d(TGG) [104]. Longer oligonucleotides d(TTGG) and d(CTGGCTCA) are platinated slightly faster on the 5′-G than on the 3′-G. Although $[Pt(NH_3)_3(H_2O)]^{2+}$ reacts slower than the *cis* derivative, platination rate constants increase with increasing oligonucleotide length for both platinum complexes. Similar behavior is seen also with *cis*-$[PtCl(NH_3)(C_6H_{11})(H_2O)]^+$ [16]. On the other hand, in the 14-base oligonucleotide d(ATACATGGTACATA), *cis*-DDP preferentially platinates the 3′-G, while the diaqua derivative shows little selectivity for the 5′-G or the 3′-G in the initial platination step for both single and double strands of the oligonucleotide [105]. Besides the ligands on Pt, also the base on the 3′ side of the GG sequence may affect the selectivity of the platination step. The dichloro species preferentially platinates the 3′-G in a d(-TGGT-) sequence and the diaqua derivative the 5′-G in a d(-TGGC-) sequence [105]. In addition, the reactions of the 5′-G in a GG sequence seem to be enhanced in the duplex compared with a single strand.

Rate constants $k_{5'c}$ and $k_{3'c}$ (Scheme 3) for the chelation step of different monofunctional adducts in which the fourth ligand is a water molecule are given in Table 5. With *cis*-$[Pt(NH_3)_2(H_2O)_2]^{2+}$, the ratio between the rates of cyclization of the 3′- and 5′-monoadducts increases from 1.4 {d(GG)} to 3.3 {d(CTGGCTCA)} [104]. With longer sequences (single-stranded and double-stranded), the preference of 3′-cyclization is even higher [105]. However, increasing the sequence length makes both chelation steps significantly slower. The increased chelation selectivity and the rate retardation of the chelation step may result from conformational changes, because single-stranded oligonucleotides adopt more B-DNA-like structure with increasing chain length [105].

With species having Cl^- as the fourth ligand, chelation rate constants are in the range $2 \times 10^{-6} - 1 \times 10^{-4}\ s^{-1}$ [8,106,115]. Most of these values are rather close to those measured for the Cl^- hydrolysis step for simple Pt(II)–nucleobase complexes of both *cis* and *trans* geometry (Section 2.2) This indicates that the hydrolysis of Cl^- ligand probably is the rate-determining step in the formation of bisadducts from monoadducts having Cl^- as the fourth ligand, as pointed out earlier [8]. However, in some cases the chelation rate constants for the Cl^- and aqua monoadducts are comparable [105,115]. In double-stranded chain, in particular, the aquation step may depend on adjacent bases and/or the incoming nucleobase (G) may directly attack the chloro-intermediate [105].

In oligonucleotides containing a GG sequence, platination of both guanine bases may also occur even in stoichiometric conditions [116]. It has been shown that the binding of $[Pt(NH_3)_3(H_2O)]^{2+}$ cations favors the 5′-G site over the 3′-G site by a factor of about 2–3 in d(CTGG) and d(CTGGCTCA) in both steps, although the affinity of both oligonucleotides toward the second $[Pt(NH_3)_3(H_2O)]^{2+}$ ion is lowered. However,

Table 5. First-Order Rate Constants for Conversion of Monofunctional Adducts of *cis*-$[Pt(NH_3)_2(\text{-G-})(H_2O)]$ into Bifunctional Adducts in Various Oligonucleotides

Platinated oligonucleotide[a]	$k_{5'c}/10^{-4}\ s^{-1}$	$k_{3'c}/10^{-4}\ s^{-1}$	*Exp. cond. [ref.]*
AG		1.5	[b] [112]
		1.44	[c] [113]
GA	4.5		[b] [112]
d(GG)	15	21	[d] [104]
d(TGG)	12.7	17.6	[d] [104]
d(GTG)		4.4	[e] [114]
d(TTGG)	13	39	[d] [104]
d(CTGGCTCA)	10	33	[d] [104]
d(ATACATGGTACATA)ss[f]	0.86[g]; 0.60[h]	4.4[g]; 2.8[h]	[i] [105]
d(ATACATGGTACATA)ds[j]	0.49[h]	2.5[h]	[i] [105]

Notes: [a]Direction of the chain 5′→3′, see Scheme 2. Charges are omitted for clarity.
[b]T = 293.2 K, pH 5.2, I not specified.
[c]T = 291.5 K, pH 3.3, I not specified.
[d]T = 293.2 K, pH 4.4, I = 0.1 M ($NaClO_4$).
[e]T = 310.2 K, pH 6.27–6.43, I not specified.
[f]Single-stranded chain.
[g]From NMR data.
[h]From HPLC data.
[i]T = 288 K, pH 4.8, I = 0.1 M ($NaClO_4$).
[j]Double-stranded chain.

increasing the eletrolyte concentration of the medium favors the second platination step. It has been suggested that the probable rate retardation effect of the first $Pt(NH_3)_3(H_2O)]^{2+}$ cation is partially compensated in the second step by lowered screening effect of the electrolyte, in agreement with other studies [16,116].

6. FINAL REMARKS

Considerable progress has been made in understanding the chemical principles in Pt–nucleic acid interactions since the discovery of Pt antitumor drugs. At the same time, however, new questions have been raised upon development of novel drugs that violate the early structure–activity relationships. A common feature for various Pt drugs is that their initial binding to nucleic acid fragments seems to be controlled by the hydrolysis of the leaving group. It appears that in the action of Pt drugs the first binding target is the G–N7 site because of favorable electronic and steric effects as well as interligand interactions associated with this binding mode. In larger fragments the preference of certain G residues is poorly understood and may depend on local effects such as base sequence and denaturation, and on the structure of the Pt drug. Factors affecting the intrinsic reactivity of the novel drugs, in particular, are at present just emerging.

The conversion of the monofunctional adducts into bifunctional lesions depends drastically on the structure of the Pt drug. Obviously, Pt compounds exhibiting *trans* geometry form different bisadducts than cisplatin and hence, a different spectrum of antitumor activity is expected. Mechanistically, the formation and possible isomerization of bisadducts are not well understood. The assumption that hydrolysis of the second leaving group controls the formation of bisadduct may be an oversimplification. Studies with model compounds as well as with oligonucleotides have indicated that a certain nucleobase may be a powerful nucleophile toward Pt(II) if spatially in a correct position. Unfortunately, our knowledge on these interactions is at present very limited.

ACKNOWLEDGMENTS

This work was supported in part by the University Foundation of Turku. The support of the colleagues in COST D8 working groups is gratefully acknowledged.

NOTED ADDED IN PROOF

The cytotoxicity of certain Pt^{IV} compounds against tumor cells may be potentiated by visible light. A very recent report shows that stereospecific reactions between a diiododiamine-Pt^{IV} complex and 5′-GMP can be induced by visible light and that photoactivation can be controlled by the axial ligands [117]. Another interesting observation shows that migration of coordinated Pt^{II}(dien) from the endocyclic to exocyclic nitrogen readily occurs in 9-methyladenine under alkaline conditions [118]. Importantly though, this easy migration reaction demonstrates the nucleophilic power of a nitrogen atom close to the metal center and may have important implication as regards the Pt-DNA rearrangements mentioned above. A very recent book describes in a fascinating way the story of

Cisplatin, the parent Pt anticancer drug, by bringing together various aspects of chemistry, biochemistry, biology, pharmacology and medicine relevant to *cis*-$[PtCl_2(NH_3)_2]$ [119].

REFERENCES

[1] See: Kaufmann, G.B., J. Chem. Educ., 26 (1957) 521.

[2] Rosenberg, B., Van Camp, L., Grimley, E.B. and Thomson, A. J., J. Biol. Chem., 242 (1967) 1347.

[3] Rosenberg, B., Van Camp, L., Trosko, J.E. and Mansour, V.H., Nature, 222 (1969) 385.

[4] Bloemink, M.J. and Reedijk, J., in Sigel, A. and Sigel, H. (eds.), *Metal Ions in Biological Systems*, Vol. 32, Marcel Dekker, New York, 1996, p. 641.

[5] Farrell, N., in Sigel, A. and Sigel, H. (eds.), *Metal Ions in Biological Systems*, Vol. 32, Marcel Dekker, New York, 1996, p. 603.

[6] Kelland, L.R., Barnard, C.F.J., Evans, I.G., Murrer, B.A., Theobald, B.R.C., Wyer, S.B., Goddard, P.M., Jones, M., Valenti, M., Bryant, A., Rogers, P.M. and Harrap, K.R., J. Med. Chem., 38 (1995) 3016.

[7] Lippert, B., Progr. Inorg. Chem., 37 (1989) 1.

[8] Bruhn, S.L., Toney, J.H. and Lippard, S.J., Progr. Inorg. Chem., 38 (1990) 477.

[9] Reedijk, J., J. Chem. Soc., Chem. Commun. (1996) 801.

[10] Whitehead, J.P. and Lippard, S.J., in Sigel, A. and Sigel, H. (eds.), *Metal Ions in Biological Systems*, Vol. 32, Marcel Dekker, New York, 1996, p. 687.

[11] Lippert, B., in Sigel, A. and Sigel, H. (eds.), *Metal Ions in Biological Systems*, Vol. 33, Marcel Dekker, New York, 1996, p. 105.

[12] Boudvillain, M., Dalbiès, R. and Leng, M., in Sigel, A. and Sigel, H. (eds.), *Metal Ions in Biological Systems*, Vol. 33, Marcel Dekker, New York, 1996, p. 87.

[13] Hartwig, J.F. and Lippard, S.J., J. Am. Chem. Soc., 114 (1992) 5646.

[14] Basolo, F. and Pearson, R., *Mechanisms of Inorganic Reactions*, Wiley, New York, 1967.

[15] Arpalahti, J., in Sigel, A. and Sigel, H. (eds.), *Metal Ions in Biological Systems*, Vol. 32, Marcel Dekker, New York, 1996, p. 379.

[16] Elmroth, S.K.C. and Lippard, S.J., Inorg. Chem., 34 (1995) 5234.

[17] Jestin, J.-L., Chottard, J.-C., Frey, U., Laurenczy, G. and Merbach, A. E., Inorg. Chem., 33 (1994) 4277.

[18] Canovese, L., Cattalini, L., Chessa, G. and Tobe, M.L., J. Chem. Soc., Dalton Trans. (1988) 2135.

[19] Knox, R.J., Friedlos, F., Lydall, D.A. and Roberts, J.J., Cancer Res., 46 (1986) 1972.

[20] Arpalahti, J., Sillanpää, R., Barnham, K.J. and Sadler, P.J., Acta Chem. Scand., 50 (1996) 181.

[21] Martin, R.B., in Lippard, S.J. (ed.), *Platinum, Gold and Other Metal Chemotherapeutic Agents*, ACS Symposium Series 209, American Chemical Society, Washington DC, 1983, p. 231.

[22] Höpp, M., Erxleben, A., Rombeck, I. and Lippert, B., Inorg. Chem., 35 (1996), 397.

[23] Pullman, A. and Pullman, B., Q. Rev. Biophys., 14 (1981) 289.
[24] Metzger, S., Erxleben, A. and Lippert, B., J. Biol. Inorg. Chem., 2 (1997) 256.
[25] Martin, R.B., Acc. Chem. Res., 18 (1985) 32.
[26] Arpalahti, J., Klika, K.D., Sillanpää, R. and Kivekäs, R., J. Chem. Soc., Dalton Trans. (1998) 1397.
[27] Meiser, C., Song, B., Freisinger, E., Peilert, M., Sigel, H. and Lippert, B., Chem. Eur. J., 3 (1997) 388.
[28] Lowe, G. and Vilaivan, T., J. Chem. Soc., Perkin Trans. 1 (1996) 1499.
[29] Trovó, G., Bandoli, G., Nicolini, M. and Longato, B., Inorg. Chim. Acta, 211 (1993) 95.
[30] Schenetti, L., Mucci, A. and Longato, B., J. Chem. Soc., Dalton Trans. (1996) 299.
[31] Mikola, M. and Arpalahti, J., Inorg. Chem., 33 (1994) 4439.
[32] Habtemariam, A. and Sadler, P.J., J. Chem. Soc., Chem. Commun. (1996) 1785.
[33] Margiotta, N., Habtemariam, A. and Sadler, P.J., Angew. Chem. Int. Ed. Engl., 36 (1997) 1185.
[34] Mikola, M., Oksman, P. and Arpalahti, J., J. Chem. Soc., Dalton Trans. (1996) 3101.
[35] Green, M., Garner, M. and Orton, D.M., Transition Met. Chem., 17 (1992) 164.
[36] Yuriev, E. and Orbell, J.D., Inorg. Chem., 35 (1996) 7914.
[37] Barnham, K.J., Bauer, C.J., Djuran, M.I., Mazid, M.A., Rau, T. and Sadler, P.J., Inorg. Chem., 34 (1995) 2826.
[38] Bau, R. and Gellert, R.W., Biochimie, 60 (1978) 1040.
[39] Guo, Z., Sadler, P.J. and Zang, E., J. Chem. Soc., Chem. Commun. (1997) 27.
[40] Arpalahti, J., Inorg. Chem., 29 (1990) 4598.
[41] Mikola, M. and Arpalahti, J., Inorg. Chem., 35 (1996) 7556.
[42] Roat, R.M., Jerardi, M.J., Kopay, C.B., Heath, D.C., Clark, J.A., DeMars, J.A., Weaver, J.M., Bezemer, E. and Reedijk, J., J. Chem. Soc., Dalton Trans. (1997) 3615.
[43] Bose, R.N. and Weaver, E.L., J. Chem. Soc., Dalton Trans. (1997) 1797.
[44] Arpalahti, J. and Lehikoinen, P., Inorg. Chem., 29 (1990) 2564.
[45] Kuusela, S. and Lönnberg, H., in Sigel, A. and Sigel, H. (eds.), *Metal Ions in Biological Systems*, Vol. 32, Marcel Dekker, New York, 1996, p. 271.
[46] Pichierri, F., Holthenrich, D., Zangrando, E., Lippert, B. and Randaccio, L., J. Biol. Inorg. Chem., 1 (1996) 439.
[47] Lippert, B., J. Chem. Soc., Dalton Trans. (1997) 3971.
[48] Sigel, R.O., Freisinger, E. and Lippert, B., J. Chem. Soc., Chem. Commun. (1998) 219.
[49] Barnham, K.J., Djuran, M.I., Murdoch, P. del S., Ranford, J.D. and Sadler, P. J., Inorg. Chem., 35 (1996) 1065.
[50] Barnham, K.J., Djuran, M.I., Murdoch, P. del S., Ranford, J.D. and Sadler, P. J., J. Chem. Soc., Dalton Trans. (1995) 3721.
[51] Barnham, K.J., Djuran, M.I., Murdoch, P. del S. and Sadler, P.J., J. Chem. Soc., Chem. Commun. (1994) 721.
[52] Djuran, M.I., Lempers, E.L.M. and Reedijk, J., Inorg. Chem., 30 (1991) 2648.
[53] Guo, Z., Chen, Y., Zang, E. and Sadler, P., J. Chem. Soc., Dalton Trans. (1997) 4107.

[54] van Boom, S.S.G.E. and Reedijk, J., J. Chem. Soc., Chem. Commun. (1993) 1397.
[55] Fröhling, C.D.W. and Sheldrick, W.S., J. Chem. Soc., Chem. Commun. (1997) 1737.
[56] Teuben, J.-M., van Boom, S.S.G.E. and Reedijk, J., J. Chem. Soc., Dalton Trans. (1997) 3979.
[57] Schmülling, M., Lippert, B. and van Eldik, R., Inorg. Chem., 33 (1994) 3276.
[58] Raudaschl-Sieber, G. and Lippert, B., Inorg. Chem., 24 (1985) 2426.
[59] Frommer, G. and Lippert, B., Inorg. Chem., 29 (1990) 3259.
[60] Beaty, J.A. and Jones, M.M., Inorg. Chem., 31 (1992) 2547.
[61] Mikola, M., Klika, K.D. and Arpalahti, J. Unpublished results.
[62] Jones, M.M. and Beaty, J.A., Inorg. Chem., 30 (1991) 1584.
[63] Mikola, M., Klika, K.D., Hakala, A. and Arpalahti, J. Inorg. Chem., 38 (1999) 571.
[64] Mahal, G., van Eldik, R., Roodt, A. and Leipoldt, J.G., Inorg. Chim. Acta, 156 (1989) 29.
[65] Bradley, L.J.N., Yarema, K.J., Lippard, S.J. and Essigmann, J.M., Biochemistry, 32 (1993) 5024.
[66] Yarema, K.J., Lippard, S.J. and Essigmann, J.M., Nucleic Acids Res., 23 (1995) 4066.
[67] Takahara, P., Frederick, C.A. and Lippard, S.J., J. Am. Chem. Soc., 118 (1996) 12309.
[68] Augé, P. and Kozelka, J., Transition Met. Chem., 22 (1997) 91.
[69] Yang, D., van Boom, S.S.G.E., Reedijk, J., van Boom, J.H. and Wang, A. H.-J., Biochemistry, 34 (1995) 12912.
[70] Fouchet, M.-H., Guittet, E., Cognet, J.A.H., Kozelka, J., Gauthier, C., Le Bret, M., Zimmermann, K. and Chottard, J.-C., J. Biol. Inorg. Chem., 2 (1997) 83.
[71] Elizondo-Riojas, M.-A., Gonnet, F., Auge-Barrere-Mazouat, P., Allain, F., Berges, J., Attias, R., Chottard, J.-C. and Kozelka J., in Banci, L. and Comba, P. (eds.), *Molecular Modeling and Dynamics of Bioinorganic Systems*, Kluver Academic Publishers, 1997, p. 131.
[72] Schröder G., Kozelka, J., Sabat, M., Fouchet, M.-H., Beyerle-Pfnür, R. and Lippert, B., Inorg. Chem., 35 (1996) 1647.
[73] Schröder G., Sabat, M., Baxter, I., Kozelka, J. Lippert, B., Inorg. Chem., 36 (1997) 490.
[74] Pérez, C., Leng, M. and Malinge, J.-M., Nucleic Acids Res., 25 (1997) 4.
[75] Paquet F., Pérez, C., Leng, M., Lancelot, G. and Malinge, J.-M., J. Biomol. Struct. Dyn., 14 (1996) 67.
[76] Vrána, O., Boudný, V. and Brabec, V., Nucleic Acids. Res., 24 (1996) 3918.
[77] Payet, D. and Leng M., Metal-Based Drugs, 2 (1995) 137.
[78] Brabec, V., Boudný, V. and Balcarová, Z., Biochemistry, 33 (1994) 1316.
[79] Sip, M., Schwartz, A., Vovelle, F., Ptak, M. and Leng, M., Biochemistry, 31 (1992) 2508.
[80] Žaludová, R., Kleinwächter, V. and Brabec, V., Biophys. Chem., 60 (1996) 135.
[81] Boudvillain, M., Dalbiès, R., Aussourd, C. and Leng, M., Nucleic Acids Res., 23 (1995) 2381.

[82] Prévost, C., Boudvillain, M., Beudaert, P., Leng, M., Lavery, R. and Vovelle, F., J. Biomol. Struct. Dyn., 14 (1997) 703.
[83] Boudvillain, M., Guérin, M., Dalbiès, R., Saison-Behmoaras, T. and Leng, M., Biochemistry, 36 (1997) 2925.
[84] Colombier, C., Boudvillain, M. and Leng, M., Antisense Nucleic Acid Drug Devel., 7 (1997) 397.
[85] Colombier, C., Lippert, B. and Leng, M., Nucleic Acids Res., 24 (1996) 4519.
[86] Manchanda, R., Dunham, S.U. and Lippard, S.J., J. Am. Chem. Soc., 118 (1996) 5144.
[87] Schliepe, J., Berghoff, U., Lippert, B. and Cech, D., Angew. Chem. Int. Ed. Engl., 35 (1996) 646.
[88] Coluccia, M., Boccarelli, A., Mariggio, M. A., Cardellicchio, N., Caputo, P., Intini, F.P. and Natile, G., Chem.-Biol. Interact., 98 (1995) 251.
[89] Goddard, P.M., Orr, R.M., Valenti, M.R., Barnard, C.F.J., Murrer, B. A., Kelland, L.R. and Harrap, K.R., Anticancer Res., 16 (1996) 33.
[90] Brabec, V., Vrána, O., Nováková, O., Kleinwächter, V., Intini, F.P., Coluccia, M. and Natile, G., Nucleic Acids Res., 24 (1996) 336.
[91] Žaludová, R., Žákovská, A., Kašparkova, J., Balcarová, Z., Vrána, O., Coluccia, M., Natile, G. and Brabec, V., Mol. Pharmacol., 52 (1997) 354.
[92] Žaludová, R., Natile, G. and Brabec, V., Anti-Cancer Drug Design, 12 (1997) 295.
[93] Kašparkova, J., Mellish, K.J., Qu, Y., Brabec, V. and Farrell, N., Biochemistry, 35 (1996) 16705.
[94] Qu, Y., Bloemink, M.J., Reedijk, J., Hambley, T.W. and Farrell, N., J. Am. Chem. Soc., 118 (1996) 9307.
[95] Mellish, K., Qu, Y., Scarsdale, N. and Farrell, N., Nucleic Acids Res., 25 (1997) 1265.
[96] Žaludová, R., Žákovská, A., Kasparková, J., Balcarová, Z., Kleinwächter, V., Vrána, O., Farrell, N. and Brabec, V., Eur. J. Biochem., 246 (1997) 508.
[97] Eastman, A., Pharmacol. Ther., 34 (1987) 155.
[98] Pendyala, L., Arakali, A.V., Sansone, P., Cowens, J.W. and Creaven, P. J., Cancer Chemother. Pharmacol., 27 (1990) 248.
[99] Nováková, O., Vrána, O., Kiseleva, V.I. and Brabec, V., Eur. J. Biochem., 228 (1995) 616.
[100] Bloemink, M.J., Engelking, H., Karentzopoulos, S., Krebs, B. and Reedijk, J., Inorg. Chem., 35 (1996) 619.
[101] Franklin, C.A., Fry, J.V. and Collins, J.G., Inorg. Chem., 35 (1996) 7541.
[102] Bernal-Méndez, E., Boudvillain, M., Gonzáles-Vílchez, F. and Leng, M., Biochemistry, 36 (1997), 7281.
[103] Gonnet, F., Kozelka, J. and Chottard, J.-C., Angew. Chem. Int. Ed. Engl., 31 (1992) 1483.
[104] Gonnet, F., Reeder, F., Kozelka, J. and Chottard, J.-C., Inorg. Chem., 35 (1996) 1653.
[105] Reeder, F., Guo, Z., Murdoch, P. del S., Corazza, A., Hambley, T. W., Berners-Price, S.J., Chottard, J.-C. and Sadler, P.J., Eur. J. Biochem., 249 (1997) 370.
[106] Barnham, K.J., Berners-Price, S.J., Frenkiel, T.A., Frey, U. and Sadler, P.J., Angew. Chem. Int. Ed. Engl., 34 (1995) 1874.

[107] Prenzler, P.D. and McFadyen, W.D., J. Inorg. Biochem., 68 (1997) 279.
[108] Kjellström, J. and Elmroth, S.K.C., J. Chem. Soc., Chem. Commun. (1997) 1701.
[109] Appleton, T.G., Berry, R.D., Davis, C.A., Hall, J.R. and Kimlin, H.A., Inorg. Chem., 23 (1984) 3514.
[110] Kozelka, J. and Barre, G., Chem. Eur. J., 3 (1997) 1405.
[111] Laoui, A., Kozelka, J., and Chottard, J.-C., Inorg. Chem., 27 (1988) 2751.
[112] Van Hemelryck, B., Girault, J.-P., Chottard, G., Valadon, P., Laoui, A. and Chottard, J.-C., Inorg. Chem., 26 (1987) 787.
[113] Alink, M., Nakahara, H., Hirano, T., Inagaki, K., Nakanishi, M., Kidani, Y. and Reedijk, J., Inorg. Chem., 30 (1991) 1236.
[114] Garner, M., Green, M., Boogaard, N. and Reedijk, J., J. Chem. Soc., Chem. Commun. (1993) 290.
[115] Berners-Price, S. J., Barnham, K. J., Frey, U. and Sadler, P. J., Chem. Eur. J., 2 (1996) 1283.
[116] Reeder, F., Kozelka, J. and Chottard, J. C., Inorg. Chem., 35 (1996) 1413.
[117] Kratochwil, N.A., Parkinson, J.A., Bednarski, P.J. and Sadler, P.J., Agnew. Chem., Int. Ed. Engl., 38 (1999) 1460.
[118] Arpalahti, J. and Klika, K.D., Eur. J. Inorg. Chem. In press.
[119] Lippert, B. (ed.), *Cisplatin—Chemistry and Biochemistry of a Leading Anticancer Drug*, Verlag Helvetica Chimica Acta, Zürich, 1999.

FUNCTIONAL MODEL COMPLEXES FOR DINUCLEAR PHOSPHOESTERASE ENZYMES

Roland Krämer and Tamás Gajda

Perspectives on Bioinorganic Chemistry
Volume 4, pages 209–240.

ISBN: 0-7623-0352-2

1. INTRODUCTION

Due to its fundamental biological role the enzymatic hydrolysis of phosphate esters has been investigated in detail. Based on their substrate specifity phosphoesterase enzymes are divided into three groups: monoesterases (phosphatases), diesterases, and triesterases. Phosphatase enzymes are important in the control of metabolic pathways (phosphorylation–dephosphorylation reactions), in protein activation/deactivation and in cellular signal transduction. Nucleases represent the most important subgroup of phosphodiesterase enzymes. The biological roles include nucleic acid degradation and modification, repair of DNA, as well as viral defense (restriction enzymes). Whereas endonucleases cleave bonds within the nucleic acid chain, exonucleases remove terminal nucleotides. Phospholipases cleave the phosphodiester bond in phospholipids such as phosphatidylcholin and are involved in the generation of second messengers. The phosphotriesterases represent a less well-characterized class of phosphoesterase enzymes. There is no important biological source of organophosphotriesters but large amounts are released into the environment as pesticides and insecticides. The discovery of phosphotriesterases is attributed to the observation that certain soil bacteria degradate organophosphotriester pesticides by rapid hydrolysis.

A growing number of protein crystal structures has provided solid evidence that in many phosphoesterase enzymes, two and sometimes even three, di- or trivalent metal ions are involved in substrate transformation. Consequently, the high catalytic efficiency is, in part, the result of a perfectly coordinated catalytic cooperation of the metal ions. Dinuclear phosphoryl transfer enzymes have been discussed thoroughly in recent reviews [1–3]. Therefore, this chapter (Section 2) only gives a brief description of enzymes for which two-metal promotion of phosphoester hydrolysis was proposed on the basis of detailed mechanistic or crystallographic studies (Table 1).

The mechanism of phosphate ester hydrolysis by hydroxide is shown in Figure 1 for a phosphodiester substrate. A S_N2 mechanism with a trigonal–bipyramidal transition state is generally accepted for the uncatalyzed cleavage of phosphodiesters and phosphotriesters by nucleophilic attack at phosphorus. In uncatalyzed phosphate monoester hydrolysis, a S_N1 mechanism with formation of a (PO_3^+) intermediate competes with the S_N2 mechanism. For alkyl phosphates, nucleophilic attack at the carbon atom is also relevant. In contrast, all enzymatic cleavage reactions of mono-, di-, and triesters seem to follow an S_N2

Table 1. Di- and Trinuclear Phosphoesterase Enzymes Discussed in this Chapter

Enzyme	*Metals*	*Substrate*	*Ref.*
Alkaline phosphatase	2 Zn^{2+} + 1 Mg^{2+}	Phosphate monoester	[4,5]
Purple acid phosphatase	Fe^{3+} + Zn^{2+}	Phosphate monoester	[6,7]
Protein phosphatase 1	Fe^{3+} + Zn^{2+}	Phosphate monoester	[8,9]
Calcineurin	Fe^{3+} + Zn^{2+}	Phosphate monoester	[10]
Inositol monophosphatase	2 M^{2+}	Inositol monophosphate	[11]
Fructose 1,6-diphosphatase	2 M^{2+}	Fructose-1,6 diphosphate	[12]
Klenow fragment	2 M^{2+}	DNA	[13–15]
Phospholipase C	3 Zn^{2+}	Phospholipids	[16]
Nuclease P1	3 Zn^{2+}	DNA, RNA	[17]
Phosphotriesterase	2 M^{2+}	Organophosphotriester	[18]

mechanism with nucleophilic attack at the phosphorus atom. At pH 7 and 25 °C, phosphate esters are very stable; the half-lifes for hydrolysis are estimated 8 years for monoalkylphosphates [19] and 200 million years for dialkylphosphates [20]. One important way to promote the hydrolysis reaction is the compensation of negative charge of the trigonal–bipyramidal transition state of the phosphoester cleavage.

Metal ions can promote hydrolytic cleavage in the following ways:

- coordination and electrostatic activation/stabilization of the substrate/transition state,
- stabilization of leaving group by metal coordination, and
- generation of a metal-hydroxide nucleophile by activation of metal-coordinated water at neutral pH.

Low molecular weight mimics for phosphoesterases are relevant from several perspectives. First, it is easier to study mechanistic features on a

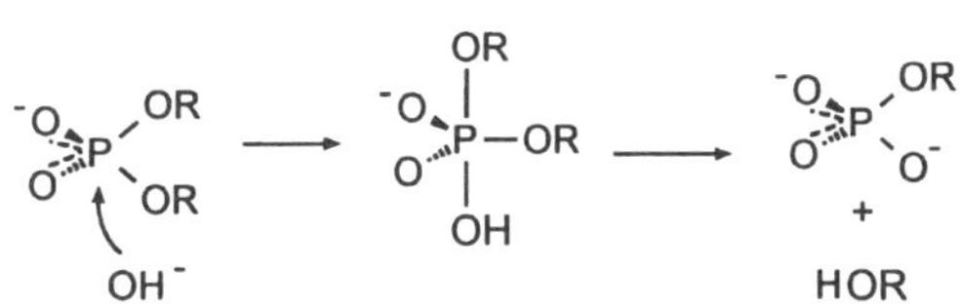

Figure 1. Generally accepted mechanism of phosphodiester cleavage by hydroxide.

simple model system than on the enzyme itself. The models may provide valuable information for understanding the chemistry involved in the active site. Second, in view of biochemical applications there is substantial interest in artificial nucleases which are smaller, more robust, and better available than the enzymes themselves. Combined with recognition agents, e.g. antisense oligucleotides, artificial nucleases are useful for the nonoxidative, sequence specific cleavage of nucleic acids and might become important tools in future biotechnology [21]. Application of such reagents as drugs for the selective control of gene expression have also been envisaged [22–24]. Third, there is a certain interest in catalysts for the mild hydrolytic decomposition of phosphotriesters in view of the detoxification of chemical warfare agents.

For the hydrolysis of phosphate esters under mild conditions, metal ions and metal complexes are the most efficient nonenzymatic reagents currently available. However, they do not reach the catalytic efficiency of enzymes, and higher reactivities are desirable in view of applications. To mimic enzymatic dinuclear sites is a strategy to generate more efficient artificial phosphoesterases.

The focus of this account is on low molecular weight metal complexes that mimic the cooperation of two metal ions in the hydrolytic cleavage of phosphate ester bonds.

2. DI- AND TRINUCLEAR PHOSPHOESTERASES

2.1 Phosphomonoesterases

Alkaline phosphatase is a nonspecific phosphomonoesterase which shows maximum activity at pH 9–10. In the *E. coli* enzyme the active site binds two Zn^{2+} ions and one Mg^{2+} ion. In 1991 Kim and Wyckoff reported an X-ray crystal structure with complexed phosphate [4]. Also, a structure of the Cd-substituted enzyme was obtained which contains a phosphorylated active site serine residue [4]. Based on these crystal structures and on many detailed mechanistic studies the reaction mechanism shown in Figure 2 was proposed. The two zinc ions (distance ca. 4 Å) are directly involved in the catalysis whereas Mg^{2+} only has a structural role. The phosphate monoester is initially bound in a 1,3-bridging fashion with a Zn1-coordinated ester oxygen atom. Nucleophilic attack of the deprotonated serine alcohol side chain results in a trigonal bipyramidal transition state with a bridging phosphate oxygen and Zn2-coordinated Ser-oxygen. The alcoholate group of the phosphomonoester

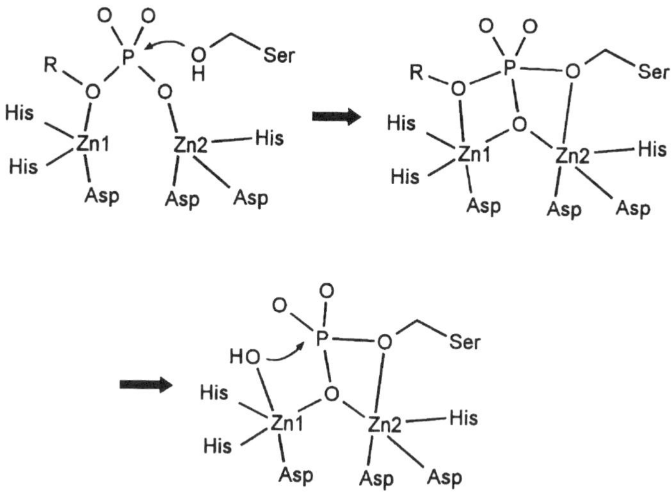

Figure 2. Reaction mechanism of alkaline phosphatase.

is released. In a second step, the phosphorylated serine intermediate is cleaved in a similar fashion by Zn1-coordinated hydroxide, resulting in the formation of phosphate. The transition state of the reaction is further stabilized by hydrogen bonding with an Arg-guanidinium group (not shown in Figure 2).

Purple acid phosphatases occur in bacteria, plants and animals and hydrolyze aryl phosphate monoesters and phosphoserine residues of phosphoproteins, although their biological role is not totally clear. In 1995 a crystal structure of kidney bean purple acid phosphatase was obtained. In the native enzyme, a Fe^{3+} and a Zn^{2+} ion with a metal–metal distance of 3.1 Å are bridged by a single oxygen atom of Asp-carboxylate and by a hydroxide ion (Figure 3). According to the proposed mechanism [6,25,26] the phosphate monoester is coordinated to the Zn^{2+} ion and an Fe^{3+}-bound hydroxide nucleophilically attacks the phosphorus atom of the substrate. Two His-imidazolium groups provide additional electrostatic activation by hydrogen bonding with the substrate (not shown in Figure 3).

Ser/Thr-protein phosphatases are ubiquitous enzymes which constitute the catalytic domains of multiprotein complexes. They are responsible for the dephosphorylation of a range of phosphoproteins. Several protein phosphatases have been characterized by X-ray crystallography and display an active site structure similar to purple acid phosphatase.

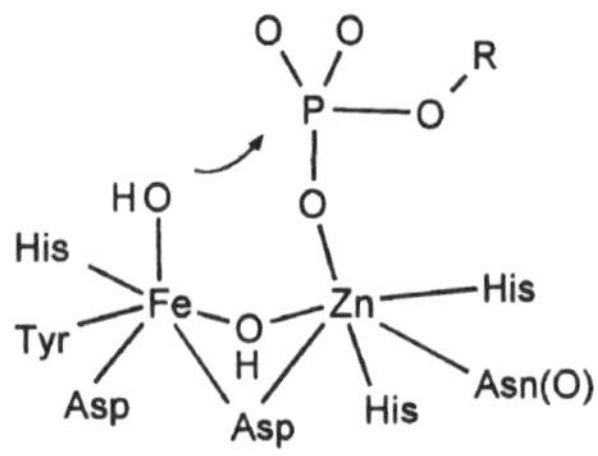

Figure 3. Reaction mechanism proposed for purple acid phosphatase.

For protein phosphatase 1 from rabbit muscle the best crystallographic resolution (2.1 Å) is available [8]. *In vivo* it is probably a Fe^{3+}-Zn^{2+} enzyme [9]. The metal–metal distance is about 3.3 Å. In spite of the structural analogies to purple acid phosphatase a different mechanism is proposed for protein phosphatase 1. The phosphate monoester is bound to both metal ions in a 1,3-bridging fashion (Figure 4). The nucleophile could either be a hydroxide ion that bridges both metals (Figure 4a) or a monodentate hydroxide at Fe (Figure 4b). Again, a NH-acidic His-imidazolium supports catalysis by activation of the substrate (not shown in Figure 4). The same mechanisms are proposed for other proteinphosphatases such as calcineurin (*in vivo* Fe^{3+}/Zn^{2+}) [10].

Inositol monophosphatase catalyzes the hydrolysis of inositol-1-phosphate, inositol-4-phosphate, and various nucleoside 2′-phosphates. The enzyme has attracted considerable interest in recent years because it is believed to be an important target for lithium therapy in treatment of manic-depression. Inositol monophosphatase inhibited in the presence

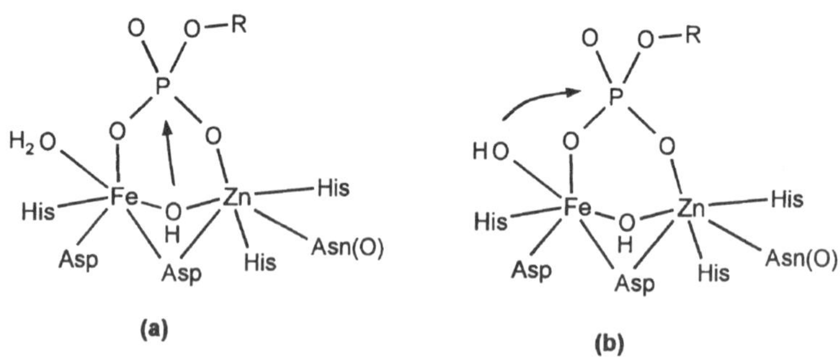

Figure 4. Possible mechanisms of phosphate monoester hydrolysis by protein phosphatase 1.

of millimolar Li^+ concentrations. Crystal structures are available with various metal ions and bound substrate analogues. In a substrate-free structure two manganese ions are located 3.9 Å apart. Mg^{2+} ions are essential for catalytic activity, but catalysis is supported by Mn^{2+}, Zn^{2+}, and Co^{2+}. The proposed reaction mechanism [27,28] is related to that of alkaline phosphatase and includes a 1,1-bridging phosphate monoester, nucleophilic attack of M2 coordinated hydroxide, and leaving group stabilization by M1 (Figure 5).

Fructose 1,6-diphosphatase hydrolyzes D-fructose 1,6-diphosphate to give D-fructose 6-phosphate and PO_4^{3-}. It is a key enzyme in the gluconeogenesis pathway. Two divalent metal ions (Mg^{2+}, Mn^{2+}, Zn^{2+}, and Co^{2+}) are involved in catalysis. In the enzyme isolated from pork kidney the metal–metal distance accounts to 3.7 Å [12]. A reaction mechanism similar to that of protein phosphatase 1 was proposed, but leaving group stabilization by metal coordination of the ester oxygen atom appears to be absent (Figure 6) [12].

2.2 Phosphodiesterases

The 3′,5′-exonuclease subunit (Klenow fragment) of *E. coli* DNA polymerase I removes bases mismatches produced by the polymerase during DNA replication. The presence of two divalent metal ions (Mg, Mn, Zn or Co, M---M ca. 4 Å) in the active site is essential for activity. Several crystallographic studies [13–15] provide good evidence for the reaction mechanism shown in Figure 7. Interestingly, the efficiency of phosphodiester cleavage seems to be based on the action of the two metal ions alone since no functional amino acid side chains are directly involved in the catalytic process. As in the case of inositol monophosphatase 1,1-bridging coordination of a phosphodiester oxygen is assumed which results in a double Lewis-acid activation of the substrate. Additionally, coordination of the 3′-ester oxygen to M2 provides leaving

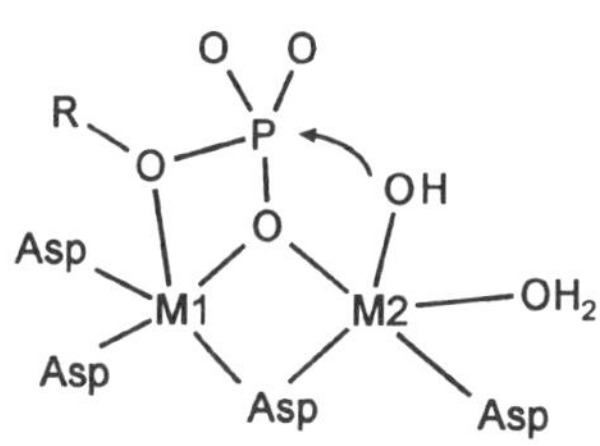

Figure 5. Proposed reaction mechanism of inositol monophosphatase.

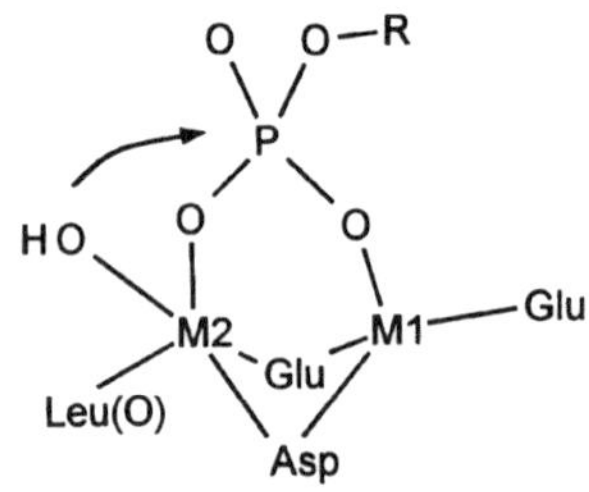

Figure 6. Proposed reaction mechanism of fructose 1,6-bisphosphatase.

group stabilization. Cleavage of the P–O bond is achieved by nucleophilic attack of M1–OH to the phosphorus atom.

Phospholipase C hydrolyzes the phosphodiester bond in phosphatidylcholin and phosphatidylinositol. By X-ray crystallography three Zn^{2+} ions were located in the active site [16] having a distorted trigonal–bipyramidal geometry in the native enzyme. Zn1 and Zn3 are 3.3 Å apart and are bridged by Asp-carboxylate. The third zinc ion Zn2 is located 4.7 Å from Zn3 and 6.0 A from Zn1. In a structure with a substrate analogue phosphonyl cholin the phosphate group coordinates all three zinc ions by the non-esterified oxygens [29]. It was proposed that the phospholipid substrates bind in the same fashion and that the nucleophile is an external water molecule which is deprotonated by Glu-carboxylate. However, molecular modeling studies are not consistent with this suggestion but favor nucleophilic attack of a Zn1–Zn3 bridging hydroxide to the Zn2-coordinated substrate.

Nuclease P1 is another trizinc enzyme which cleaves the phosphodiester bond in single-stranded RNA and DNA. Protein crystallography has revealed that the structure of the three zinc site is very similar to that

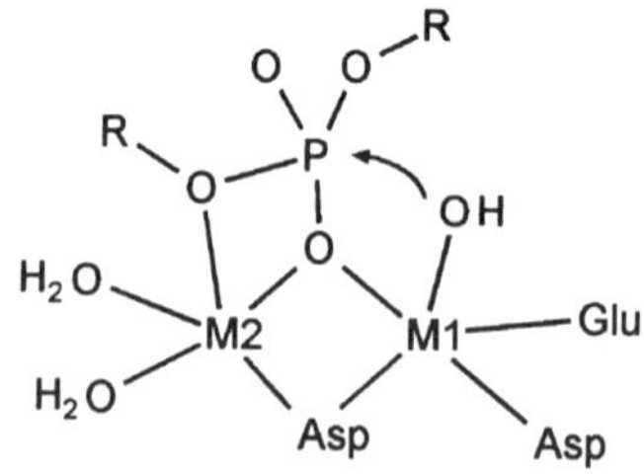

Figure 7. Reaction mechanism of the 3′,5′-exonuclease subunit DNA polymerase I.

observed in Phospholipase C [17]. A mechanism with attack of a Zn(μ-OH)Zn nucleophile to the substrate coordinated to the third Zn is favored [1], similar to a mechanistic possibility discussed for phospholipase C.

2.3 Phosphotriesterases

A phosphotriesterase isolated from the soil bacterium *Pseudomonas diminuta* is the best characterized enzyme of this type. There is evidence for the presence of two active site Zn^{2+} ions *in vivo*. A crystal structure of the dinuclear Cd^{2+} form is available in which the metal ions are bridged by a carbamylated Lys-amino group with a metal–metal distance of 3.8 Å [18]. Substrate hydrolysis follows a S_N2 type reaction and nucleophilic attack of M–OH is likely, but mechanistic details are not yet clear.

3. MODEL COMPLEXES

3.1 Substrates

The natural substrates of phosphoesterase enzymes, in particular phosphodiesters, are rather resistant to hydrolysis. For example, there are very few reports for a detectable nonenzymatic hydrolysis of linear DNA under mild conditions. Therefore, many model studies use activated *p*-nitrophenyl esters which are more reactive. Often, the trends in reactivity observed with activated substrates can be transferred to their nonactivated counterparts [30]. This is not the case, however, if leaving group stabilization is important since the pK_a of a nitrophenol leaving group is about 7 orders of magnitude smaller than the pK_a of an aliphatic alcohol. Spontaneous cleavage of the phosphodiester bond in RNA is much faster compared to DNA since the ribose 2′-OH group acts as an intramolecular auxiliary group. 2-Hydroxypropyl-*p*-nitrophenyl phosphate (HPNP), an activated RNA-analogue, releases *p*-nitrophenolate by intramolecular nucleophilic attack of the OH-group and formation of a cyclic phosphodiester.

The phosphodiester substrates used for the model studies described herein are shown in Figure 8. ApA and 2′,3′-c-UMP represent several diribonucleotides and cyclic 2′,3′-cyclic ribonucleotides.

3.2 Kinetically Inert Cobalt(III) Complexes

The cobalt(III) ion has a small size-to-charge ratio and is therefore a strong Lewis acid which strongly activates coordinated phosphate ester

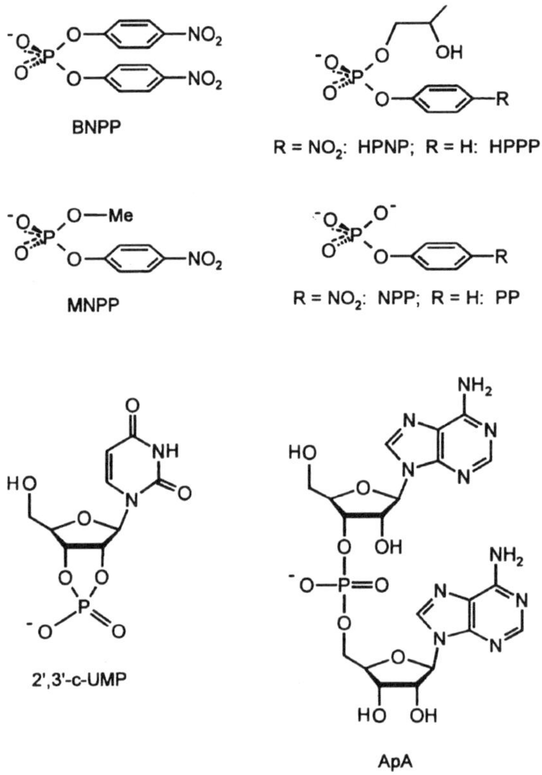

Figure 8. Substrates used in the model studies described in Section 3.

substrates. Furthermore, ligand exchange at cobalt(III) is slow. For example, hydrolysis of a Co (III)-coordinated phosphodiester results in formation of a phosphate monoester which is not readily replaced by another substrate molecule. Therefore, catalytic turnover is not observed on exchange inert cobalt(III), and these systems are not relevant to the rational design of artificial phosphoesterases that cleave phosphate ester bonds with catalytic turnover. On the other hand, studies on cobalt(III) complexes have substantially contributed to the basic understanding of metal-promoted phosphate ester hydrolysis since kinetic analysis is not complicated by rapid substrate exchange.

In 1984 Sargeson and coworkers [31] have detected that the dinuclear, *p*-nitrophenyl phosphate-bridged cobalt(III) complex **1** rapidly hydrolyzes in water at pH 10 and 25 °C ($k = 2 \times 10^{-2}\ s^{-1}$). Based on a detailed kinetic investigation the authors suggest intramolecular attack of coordinated hydroxide to the bridging phosphate ester in the ring-opened species **2**. Hydrolysis reaction is 10^7 times faster than spontaneous hydrolysis of *p*-nitrophenyl phosphate at pH 10 and 25 °C. The presence of the second cobalt center in **2** is responsible for a 26-fold increase in rate over that observed for the mononuclear complex **3** ($k = 8 \times 10^{-4}\ s^{-1}$). The mechanism is related to phosphomonoester hydrolysis by fructose 1,6-bisphosphatase (Figure 6) and protein phosphatase 1 (Figure 4).

A similar reaction mechanism was proposed by Chin et al. [32] for the hydrolysis of the biological phosphate monoester adenosine monophosphate (AMP) by the complex $[(trpn)\ Co\ (OH_2)]^{2+}$ [trpn = tris(aminopropyl)amine]. Rapid cleavage is observed only in the presence of 2 equiv metal complex. It is evident from ^{31}P NMR spectra that on coordination of 1 equiv (trpn)Co to AMP a stable four-membered chelate complex **4** is formed. The second (trpn)Co molecule may bind to another oxygen atom of the substrate (formation of **5**) and provide a Co–OH nucleophile which replaces the alkoxy group. The half-life of AMP in **5** is about 1 h at pD 5 and 25 °C.

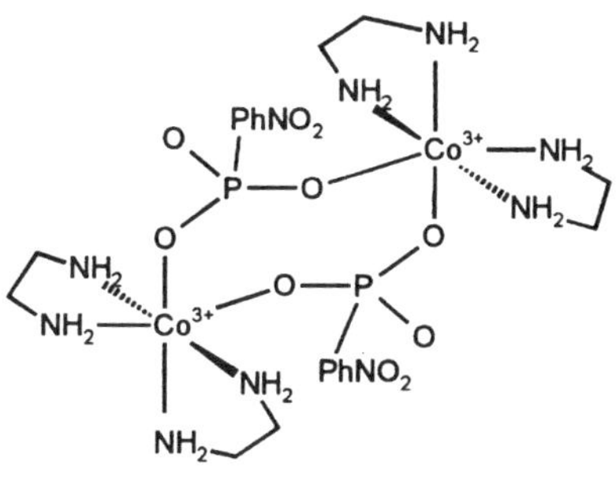

1

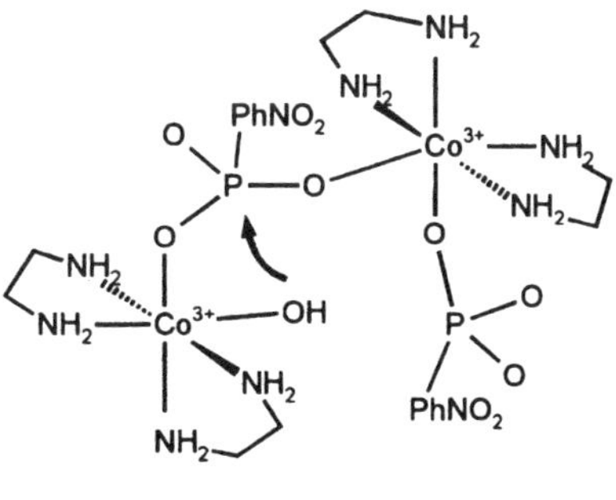

2

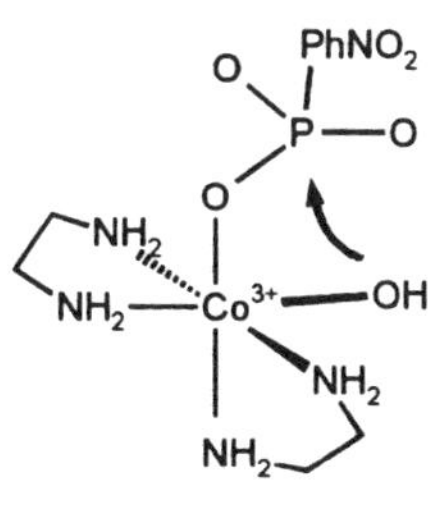

3

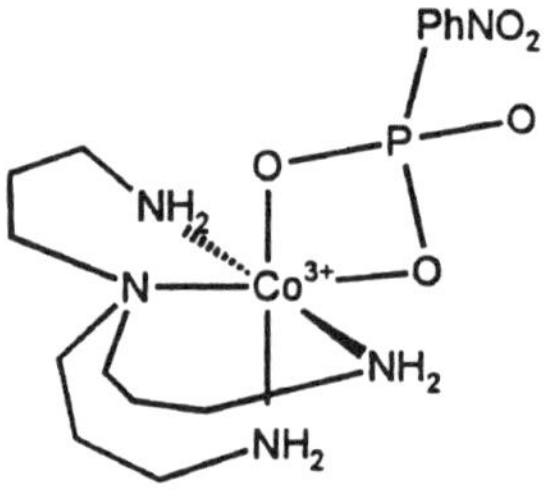

4

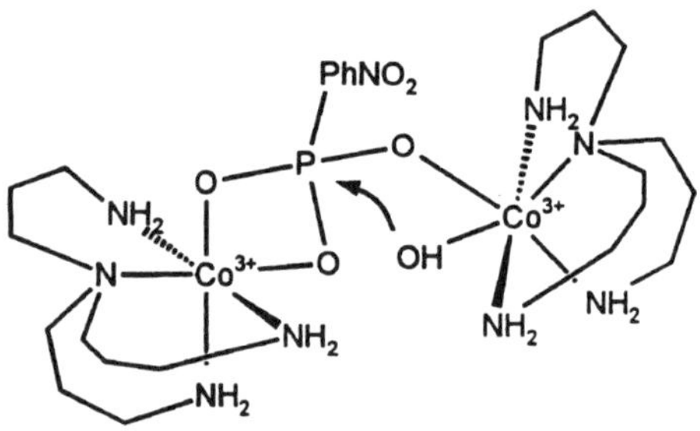

5

In Czarnik's model compound **6** both cobalt(III) ions are presented in the same molecule [33]. Two (cyclen)Co(III) units are covalently linked by a 1,4-dibutylbenzene spacer. By **6** the activated phosphodiester BNPP at pH 7.0 and 25 °C is hydrolyzed 3.2 times faster than by 2 equiv of (cyclen)Co(III) (**7**) under the same conditions. A more than 107-fold rate enhancement over the spontaneous hydrolysis of BNPP is observed. The reaction mechanism may be similar to that proposed for compound **2**.

In complex **8**, also prepared by Czarnik and coworkers [34], the two (cyclen)Co(III) units are linked by a different spacer. Models predicted that the complex is prohibited from forming an intramolecular μ-hydroxo dimer. Furthermore, the mononuclear subunits are more rigidly held compared with complex **6**. Whereas no rate increase over **7** is achieved for BNPP hydrolysis, cleavage of the monoester NPP by **8** is 10 times faster than by (cyclen)Co in twofold concentration. ^{31}P NMR data indicate that a four-membered chelate may be involved in the cleavage mechanism, as observed in the case of **5**.

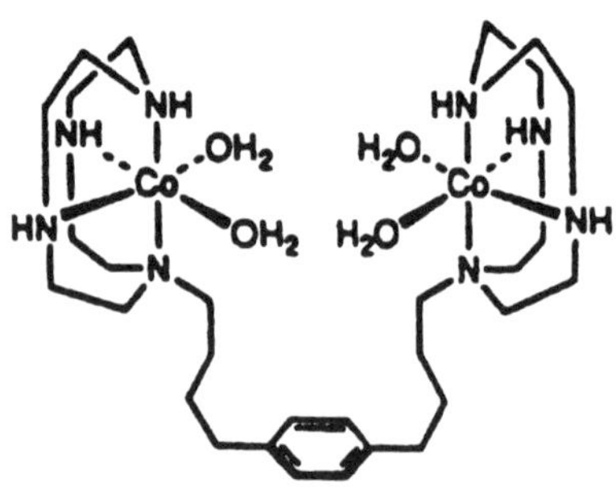

6

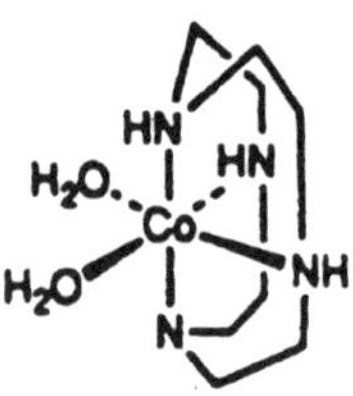

7

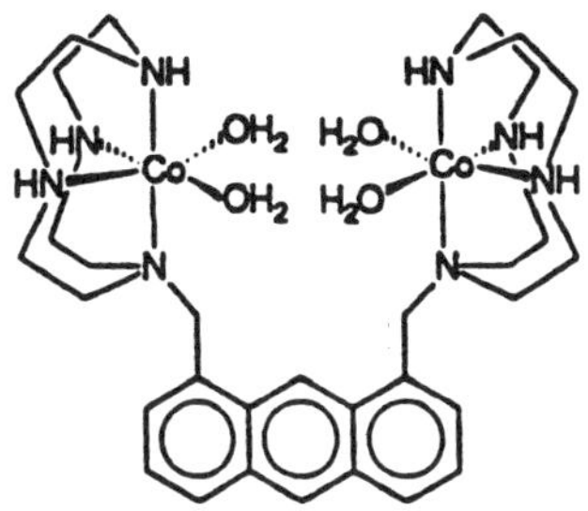

8

The highest rate acceleration in the nonenzymatic hydrolysis of a phosphate monoester was reported by Chin's group [35]. In the dinuclear cobalt(III) complex **9** the metal ions are much more rigidly preorganized than in complexes **6** and **8**. At pH 7 and 25 °C coordinated phenyl phosphate (PP) hydrolyzes 10^{11} times faster than free PP under the same conditions. There is good evidence for a reaction mechanism which has already been suggested for **2**. The higher reactivity of **9** compared to **2** may be attributed to the proximity of substrate and M–OH nucleophile.

Also, Chin's dihydroxy-bridged dicobalt(III) complex **10** provides one of the largest rate enhancements ever observed for phosphodiester hydrolysis [36]. Reaction with the activated phosphodiester methyl(p-nitrophenyl)phosphate (MNPP) yields **11**. A crystal structure is available for the analogous dimethylphosphate complex in which the Co ions are 2.9 Å apart. At pH 7 and 45 °C **11** releases *p*-nitrophenolate with $k = 0.1$ s^{-1} which corresponds to a nearly 10^{12}-fold rate enhancement over spontaneous hydrolysis of the substrate. The product of the reaction is a methylphosphate-bridged complex. Based on the pH rate profile and on

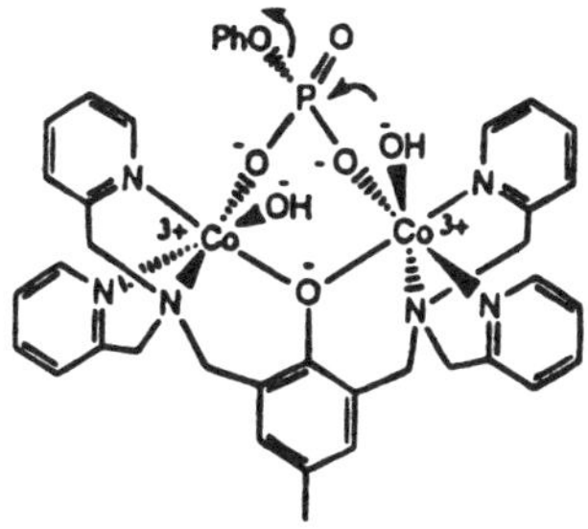

9

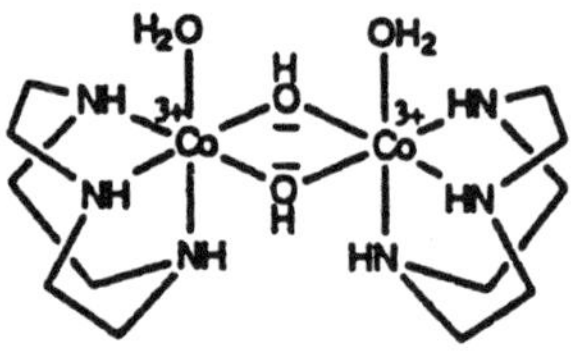

10

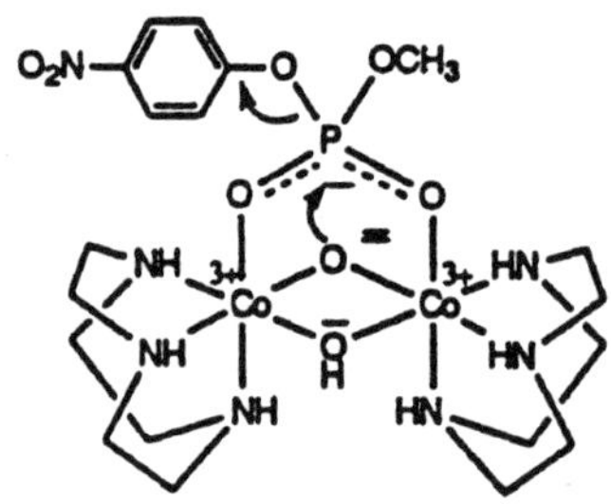

11

isotope labeling studies, an unusual nucleophilic attack of bridging oxide [O^{2-}, for Co(μ-OH)Co $pK_a = 14$ is estimated] at the phosphorus atom is proposed. This feature is related to the mechanistic proposal for phosphate monoester hydrolysis by protein phosphatase 1 (Figure 4a).

Compound **10** has also been used to quantify double Lewis acid activation by two cobalt (III) ions [37]. In **12**, the RNA analogue 2-hydroxypropyl-phenyl phosphate (HPPP) is coordinated to the dinuclear cobalt site. It is well known that in this substrate the hydroxypropyl group is an efficient intramolecular nucleophile. Release of phenol by intramolecular cyclization is much faster than the reaction by nucleophilic attack of bridging oxide, as observed in **11**. At pH >8, transesterification rate is linearly dependent on hydroxide concentration since OH^- acts as an intermolecular base for the deprotonation of the hydroxypropyl group. The second order rate constant for the hydroxide-dependent cleavage is 4×10^5 times larger than the second-order rate constant for the hydroxide-dependent spontaneous transesterification of hydroxypropyl-phenyl phosphate.

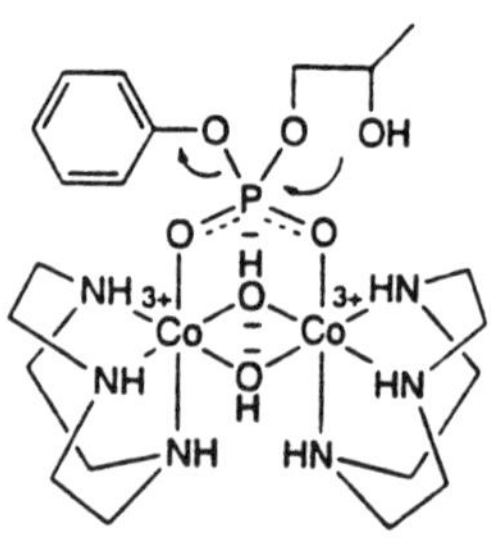

12

3.3 Kinetically Labile Complexes

In spite of the above mentioned Co(III) compounds, kinetically labile metal complexes may provide fast product/substrate exchange and some of these systems show real catalytic activity. In native dinuclear phosphatases Mg(II), Mn(II), Fe(II/III), or Zn(II) ions are present in the active centers. Although the aqua complexes of the weakest Lewis acids, Mg(II) and Mn(II), show measurable acceleration of e.g. the transesterification of 2-hydroxypropyl *p*-nitrophenyl phosphate {HPNP, [Mn(II)] = 0.004 M, k_{obs}/k_{uncat} = 73 at pH 7 and 310 K, [38]} or the hydrolysis of 3′,5′-uridyluridine (UpU) [39], only a few structural [40] but no functional phosphatase-mimicking dinuclear complexes have been reported with these metal ions.

The higher coordinating ability and Lewis acidity of Zn(II) ion in addition to the low p*K* of the metal-bound water molecule and the appearance of this metal ion in native phosphatases inspired a number of research groups to develop Zn(II)-containing dinuclear artificial phosphatases. In contrast, very few model compounds have been published to mimic the activity of Fe(III) ion in dinuclear centers of phosphatase enzymes. Cu(II) or lanthanide ions are not relevant to natural systems but their chemical properties in certain cases allow extraordinarily high acceleration of phosphate-ester hydrolysis [as much as 10^8 for copper(II) or 10^{13} for lanthanide(III) ions].

Dizinc(II) Complexes

An early attempt to achieve metal–metal cooperation on the nonenzymatic hydrolysis of a phosphate triester failed, probably due to the long and flexible linker connecting two imidazole-containing metal-binding sites [41].

The dimer complex **13** showed, for the first time for kinetically labile complexes, an enhanced activity toward the hydrolysis of the activated

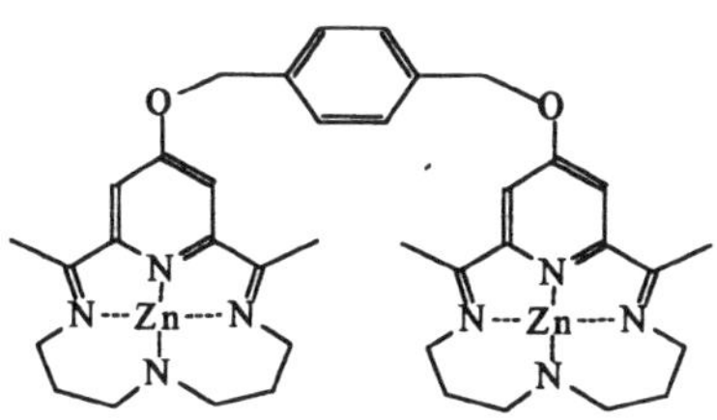

13

p-nitrophenyl diphenyl phosphate over the corresponding monomer complex [42]. The rate enhancement is, however, too modest ($k_{dimer}/2*k_{monomer} = 2.2$) to establish the exact way of cooperation of the two metals.

As an extension of this work, Chapmann and Breslow prepared [43] a series of dinucleating ligands (**14–17**) able to hold two metal ions at various distances. Their hydrolytic activity was examined (pH = 8.36 in 20% DMSO-water) toward NPP ($T = 328$ K), BNPP ($T = 328$ K), HPNP ($T = 303$ K) and UpU ($T = 314$ K). The rate enhancements caused by the plausible metal–metal cooperativity in complexes containing rather rigid spacers range between 4 and 7, while in the case of the flexible pentamethylene linker between 0.75 and 3, as compared to the corresponding monomer. The rate enhancement has also been shown to depend on the metal–metal separation induced by the different size and geometry of the spacers (**16** ca. sevenfold more active against BNPP, while 4 times less active in the hydrolysis of NPP than **14**). These behaviors have been explained by the different binding mode of the substrates to the metal centers: (1) in case of the more basic monoester the coordination of both Zn(II) ions to the phosphate is needed to achieve sufficient Lewis-acid activation while one of them provides the nucleophilic OH, and (2) in case of phosphate diesters only one zinc acts as a Lewis-acid activator (**18**). The mechanism proposed for **16** is related to that of purple acid phosphatase. The pH-rate profiles measured for **14** and **16** (with NPP and BNPP, respectively) showed bell-shaped curves in both cases. The monohydroxo form in both complexes is proposed to be the active species, while the second deprotonation produces an inactive dihydroxo species. From the pH rate profiles a first p$K \approx 8$ of Zn–OH_2 for both complexes but a very different second protonation constant (pK 7.8 and 12 for **16** and **14**, respectively) was derived. The presented pH-metric curves, on the other hand, do not really support the postulated large difference in the second pK values of complexes **14** and **16**.

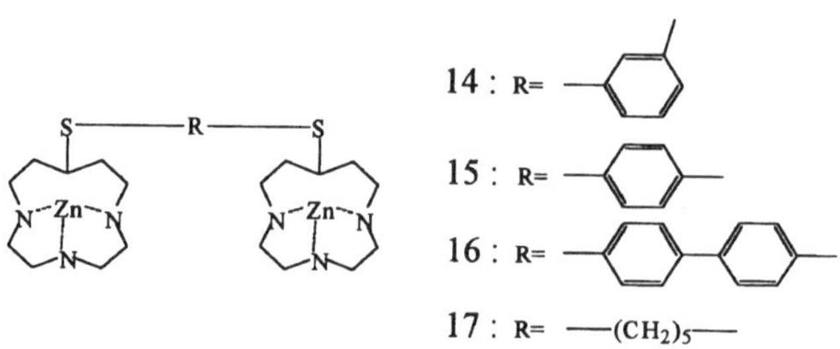

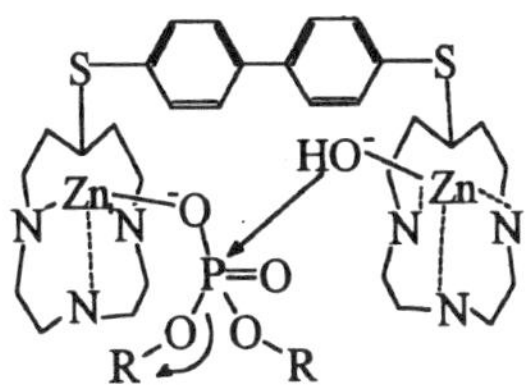

18

Recently, the Zn(II) complex of the binucleating tetraoxa-hexaaza macrocycle **19** has reported to promote the hydrolysis of BNPP [44], although catalytic turnover was not examined. The kinetic study (pH = 8–10.4, T = 308 K) indicated that the $Zn_2L(OH)_2$ complex accelerates the hydrolysis of BNPP, showing ca. fivefold rate enhancement compared to the corresponding ZnLOH complex of the mononucleating ligand **20**. At the same time, no rate enhancement by the second metal was observed on the hydrolysis of *p*-nitrophenyl-acetate, where the metal catalysts are known to act merely as general base. As a result, the authors concluded double Lewis-acid activation of the dimer complex (which may be also supported by the crystal structure of the diphenyl phosphate coordinated Zn_2L complex) and the metal-bound hydroxides provide the nucleophilic catalysis (**21**). The proposed mechanism is similar to that of several phosphomonoestarases.

Komiyama at al. have prepared two oligonuclear Zn(II) complexes (**22** and **23**) and tested their hydrolytic activity toward different diribonucleotides [45,46] (catalytic turnover was not demonstrated). The dimer and trimer structures of the active species were confirmed by measuring the hydrolytic activity as a function of Zn/L ratio, which show sharp maxima at the expected 2/1 and 3/1 ratios, respectively. The oligomer complexes have high ribonuclease activity (e.g. the hydrolysis of UpU is accelerated more than 4 and 5 orders of magnitude by **22** and **23**, respectively), whereas the effect of the monomer complex **24** was not

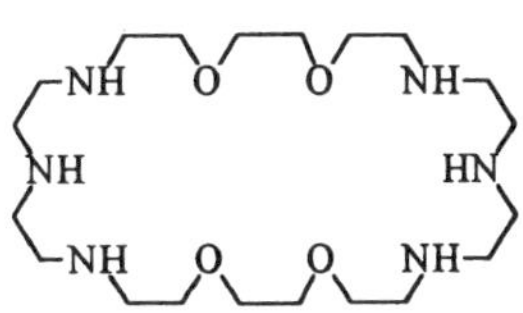

19

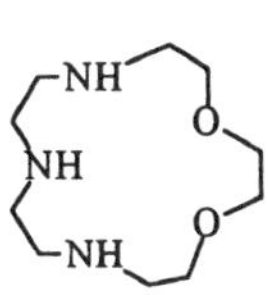

20

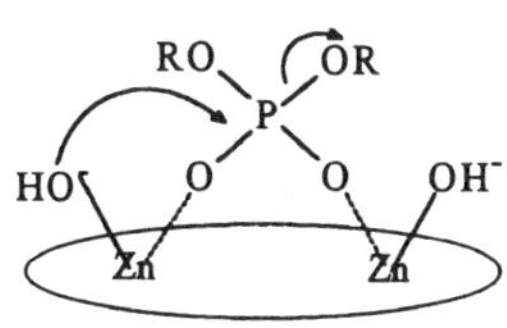

21

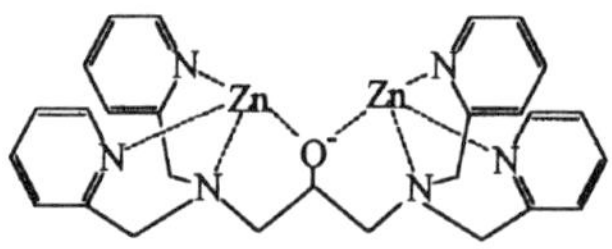

22

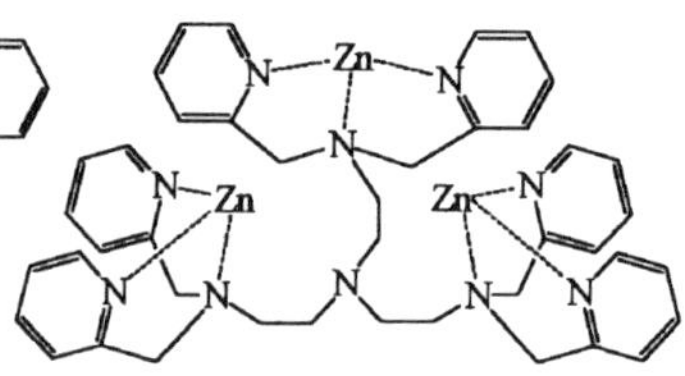

23

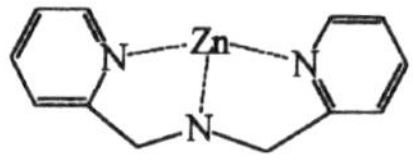

24

detectable under the conditions used ([complex] = 5 mM, pH = 7, T = 323 K). The complex **22** provides nearly the same acceleration (k_{obs} = 5–6∗10^{-5} s^{-1}) for four different diribonucleotides studied and shows low regioselectivity. The suggested roles of the metal ions are: (1) promotion the deprotonation of 2′-OH, (2) stabilization of the five-coordinated intermediate, and (3) facilitation of the removal of the 5′-O^- from the intermediate. The postulated mechanism may need further refinement since the protonation state of the active complex is unknown.

In contrast to complex **22**, **23** shows notable substrate selectivity (k_{obs} ranges between 8 and 72∗10^{-5} s^{-1} for six diribonucleotides studied) and high regioselectivity (the formation of the 3′-NMP isomer is strongly favored in all cases). The authors do not propose a mechanism for the hydrolysis, which in fact would be difficult since the three metal centers may cooperate in a number of ways. On the other hand, clarifying the mechanism would greatly help to achieve highly base- and regioselective artificial nucleases. One might speculate based on the reported crystal structure of the trinuclear Cu(II) complex of the ligand in question [47], which shows unsymmetrical (one 4N and two 3N coordinated) metal centers. Such an asymmetry may be relevant to the observed regioselectivity. The above mentioned Cu(II) trimer, beside by an oxidative mechanism, probably cleaves plasmid DNA in a hydrolytic manner, too [47].

Reinhoudt at al. have recently reported [48] the preparation of a calix[4]arene functionalized with two Zn(II) centers **25**, which is highly efficient on transesterification of HPNP. The dimer complex is reported to be 50 times more active (in 50 (v/v)% acetonitrile–water at pH = 7.4 and I = 298 K) than the corresponding monomer (**26**) which is itself 6 times more active than **27**, implying the contribution of the calix[4]arene moiety in the mechanism. The saturation kinetic experiments showed high association constant for the catalyst–substrate complex (K_{ass} =

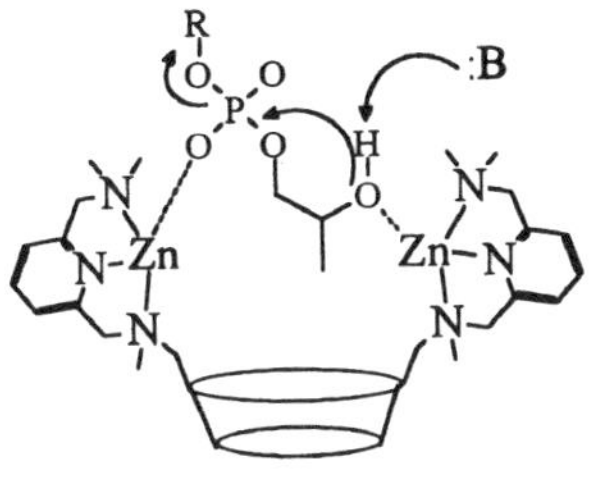

25

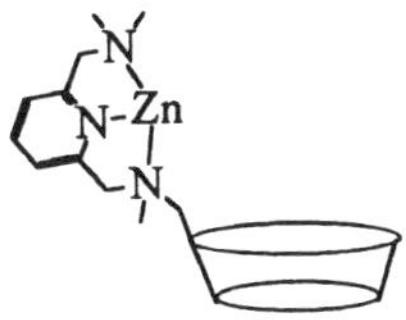

26

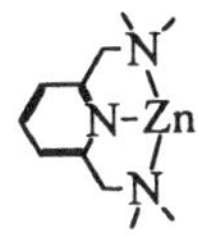

27

 = calix[4]arene

$5.5*10^4\ M^{-1}$ at pH = 7.0). This was explained by the synergetic action of the directionally preorganized two metal centers and the ability of the calix[4]arene ring to adopt optimal conformation. The pH-rate profile is bell shaped, as a result of the opposing effect of the decreasing K_{ass} and increasing k_{cat} approaching to higher pH. According to the proposed bifunctional mechanism (**25**) one of the Zn(II) ion serves as Lewis-acid activator while the other activates the nucleophilic β-OH group of the substrate. Four catalytic turnovers were detected.

In a detailed study, Kimura et al. have reported a dinuclear, alkoxide-bridged Zn(II) complex **28** of an octaazacryptand possessing unique behavior in the reaction with phosphate monoesters [49]. The crystal structure of the alkoxide-bridged dimer revealed two identical zinc centers having trigonal–bipyramidal coordination sphere. The complex selectively cleaves NPP ($k_{obs} = 7.5*10^{-6}$ at pH = 6 and T = 308 K with 5 mM complex) by the nucleophilic attack of a secondary amine. The crystal structure of the resulting intermediate **29** is also available and shows bridging metal coordination of the phosphoramide oxygens, which results in the high stability of this intermediate. Therefore, the reaction is not catalytic. The phosphoryl-transfer reaction is characterized by a bell-shaped pH-rate profile. The increase of activity with pH corresponds to the deprotonation of the $HNPP^-$ into the active NPP dianion, while the loss of activity has been assigned to the deprotonation of a secondary ammonium group in the dimer–npp associate **30**, which electrostatically disfavors binding and cleavage of the substrate. The resulting complex possesses ca. sixfold lower activity compared to the protonated one.

Mechanistically it was proposed that the substrate bridges both Zn(II) ions and replaces the apical amines in **28** while one of them attacks the phosphorus atom as an intramolecular nucleophile to perform the phosphoryl-transfer reaction (**30**). The presented observations are not re-

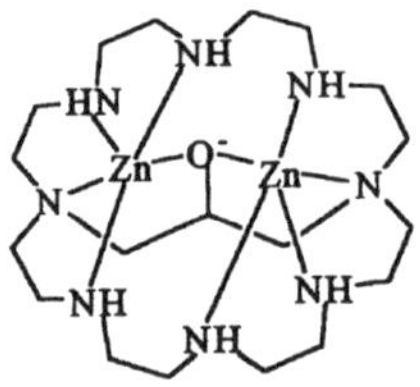

28

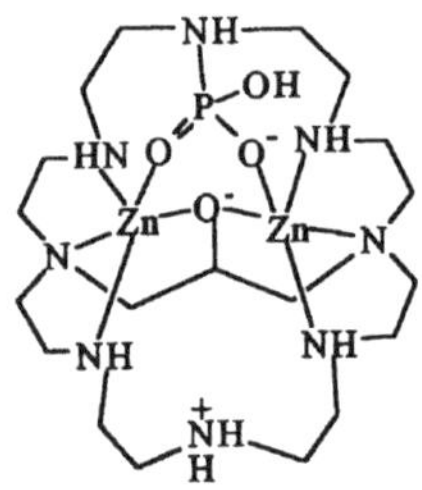

29

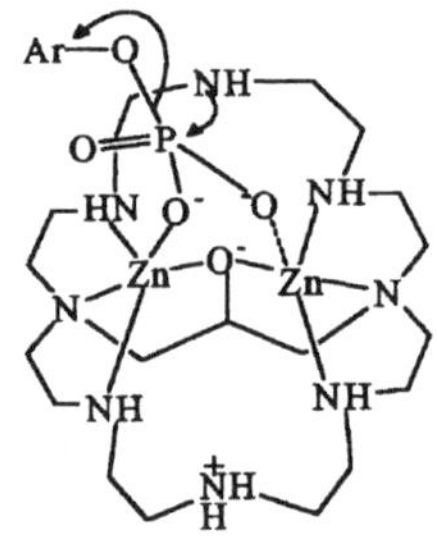

30

stricted to NPP, since analogous reactions were detected with adenosine-5′-triphosphate (ATP), too.

More recently, the Zn(II) complex of a diimidazole ligand **31** was used to mimic ribonucleases [50]. In presence of metal excess ([Zn]/[L] = 2/1), two strongly overlapped deprotonations were observed (in 65 w/w% ethanol–water) around pH = 7.3. Parallel with this process, an important increase of hydrolytic activity was observed, with a sigmoidal pH-rate profile. Both steps of spontaneous RNA hydrolysis, cyclization, and ring opening, were modeled by transesterification of HPNP (298 K) and hydrolysis of 2′,3′-cUMP (310 K). The hydrolytic efficiency was found higher toward 2′,3′-cUMP than toward HPNP (k_{obs}/k_{uncat} ca. 1400 and 10^4 with 1 mM complex at pH = 8) . Kinetic data indicated that the dinuclear complex is the reactive species. According to the postulated mechanism the complex operates with double Lewis-acid activation, while the metal-bound hydroxide (or alkoxide) serves as general base catalyst, interacting with the β-hydroxyl group. Approximately three catalytic turnovers were demonstrated.

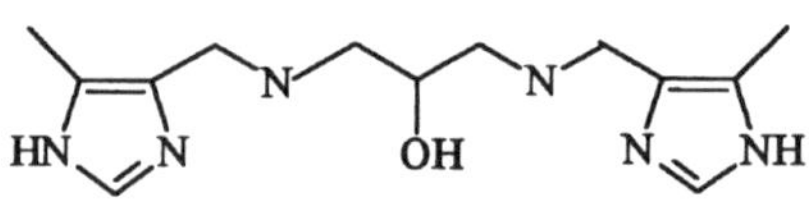

31

Dicopper(II) Complexes

The monomeric *cis*-diaqua Cu(II) complex **32** has been shown to promote efficiently the transesterification of HPNP (pH = 8 and T = 298 K) with second-order dependence on complex concentration [51]. In this respect **32** shows unique behavior, since a number of similar species are known to form inactive μ-dihydroxy-bridged dimers with increasing concentration. The pH dependence of the rate of transesterification indicated that the aqua–hydroxy form or its kinetic equivalent is the active species. In contrast to many other *cis*-diaqua complexes, in the present case formation of the monohydroxy-bridged dimer (being in equilibrium with the less active monomer) is favored for steric reasons. The dimer efficiently promotes the transesterification by a double Lewis-acid activation **33**.

Chin at al. have also demonstrated [52] notable bimetallic cooperativity with the same substrate by the Cu(II) complex **34**. The dimer complex is 26 times more active (at pH = 7 and $T = 298$ K) than the corresponding mononuclear species **35**. Based on the crystal structure of the dibenzyl phosphate bridged complex, the authors have proposed double Lewis-acid activation, as in the preceding case.

Even more efficient bimetallic cooperativity was achieved by the dinuclear complex **36** [53]. It was demonstrated to cleave 2′,3′-cAMP (298 K) and ApA (323 K) with high efficiency at pH 6, which results in 300–500-fold rate increase compared to the mononuclear complex Cu(II)-[9]aneN at pH 7.3. The pH-metric study showed two overlapped deprotonations of the metal-bound water molecules near pH 6. The observed bell-shaped pH-rate profiles indicate that the monohydroxy form is the active species. The proposed mechanism for both 2′,3′-cAMP and ApA hydrolysis consists of a double Lewis-acid activation of the substrates, while the metal-bound hydroxide acts as general base for activating the nucleophilic 2′-OH group in the case of ApA (**36a**). Based on the 1000-fold higher activity of the dinuclear complex toward 2′,3′-cAMP, the authors suggest nucleophilic catalysis of the Cu(II)–OH unit in **36b**. The latter mechanism is comparable to those of protein phosphatase 1 and fructose 1,6-diphosphatase.

Hamilton at al. have prepared two dinuclear Cu(II) complexes (**37**, **38**) and shown that these compounds are active for HPNP transesterification

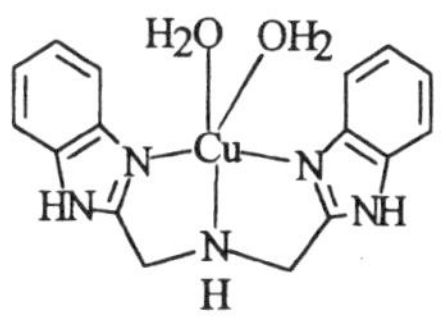

32

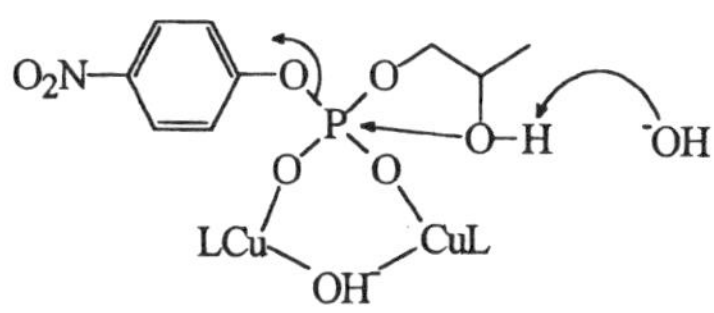

33

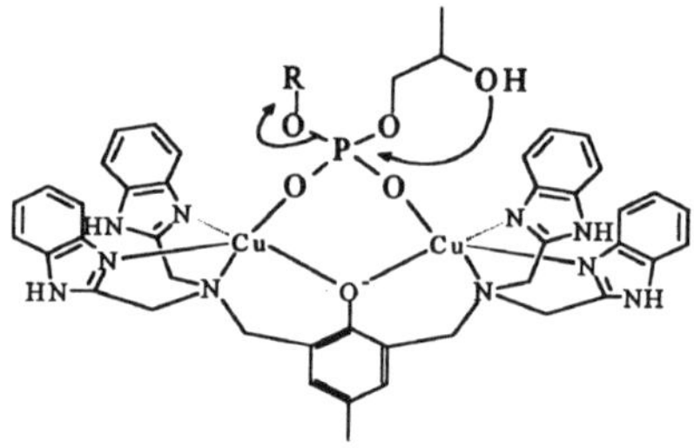

34

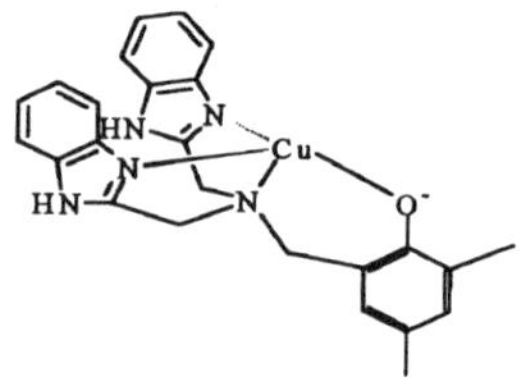

35

[54]. Their activity compared to those of the Cu(II)–terpyridine and Cu(II)–bipyridine complexes indicate notable cooperativity between the metal centers ($k_{dimer}/2*k_{monomer}$ = 18–26 at pH = 7 and ca. 10 at the pH optimum of the given complex). The pH-rate profile of both complexes shows a bell-shaped structure. Thus, the postulated double general-base catalysis for both complexes is not fully justified. In case of **38** this was explained by possible inhibition by the buffer used. While double Lewis-acid activation is proposed for **37**, single Lewis-acid activation is favored for **38**.

The above complexes have been shown to mimic the second step of RNA hydrolysis as well, i.e. the-efficient cleavage of ribonucleoside 2′,3′-cyclic monophosphates [55] with bell-shaped pH-rate profile. With these substrates **37** showed much higher bimetallic cooperativity; the $k_{dimer}/2*k_{monomer}$ ratios range between 64 and 457 for the different 2′,3′-NMPs used, while for **38** this ratio varies between 1 and 26. Since the mononuclear complexes have nearly the same activity toward the different 2′,3′-NMPs, these kinetic data indicate a notable base-selectivity of the dimer complexes. Since no correlation was observed with the size,

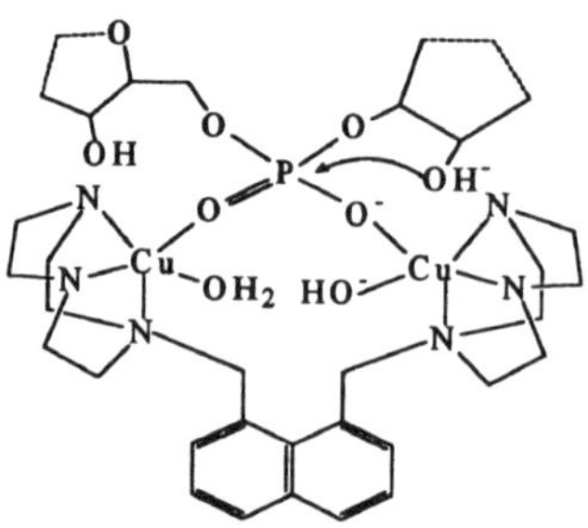

36a

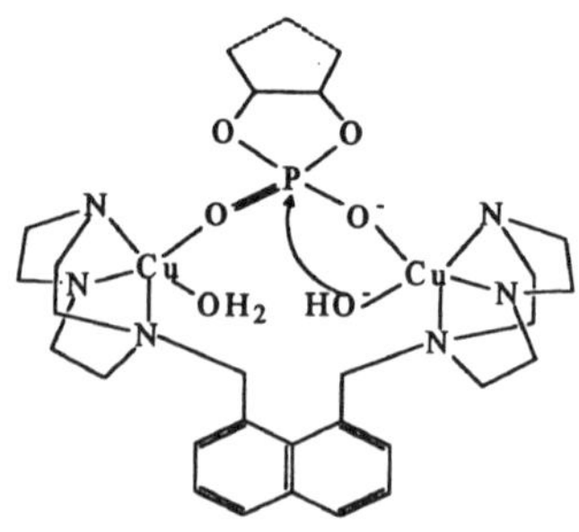

36b

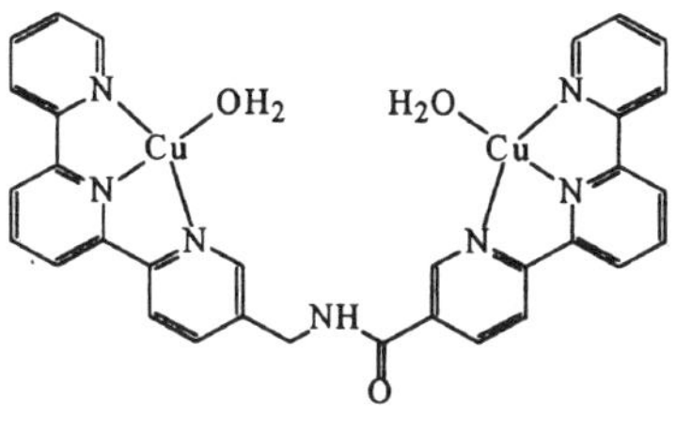

37

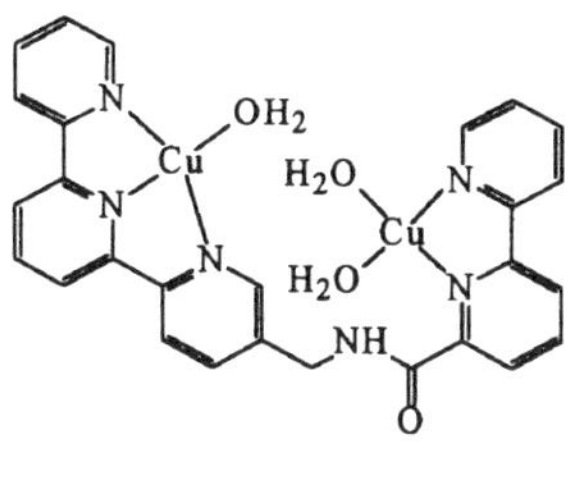

38

shape, or metal-binding ability of the base moiety, the different reactivities are probably due to the ligand–substrate interaction, e.g. H-bond with the amide group or stacking interaction of aromatic rings.

Even more interesting is the observed regioselectivity of **37**; its reaction with 2′,3′-cCMP and 2′,3′-cUMP resulted in formation of more than 90% of 2′-phosphate (3′-OH) isomer. The postulated mechanisms for **37** consists of a double Lewis-acid activation, while the metal-bound hydroxide and water act as nucleophilic catalyst and general acid, respectively (see **39**). The substrate–ligand interaction probably favors only one of the depicted substrate orientations, which may be responsible for the observed regioselectivity. Complex **38** may operate in a similar way but with single Lewis-acid activation, which would explain the lower bimetallic cooperativity and the lack of regioselectivity. Both proposed mechanisms show similarities to that of the native phosphomonoesterases (**37**: protein phosphatase 1 and fructose 1,6-diphosphatase, **38**: purple acid phosphatase).

Diiron(III) Complexes

The first reported diiron complex possessing nuclease activity actually did not operate by a true hydrolytic mechanism, although the products

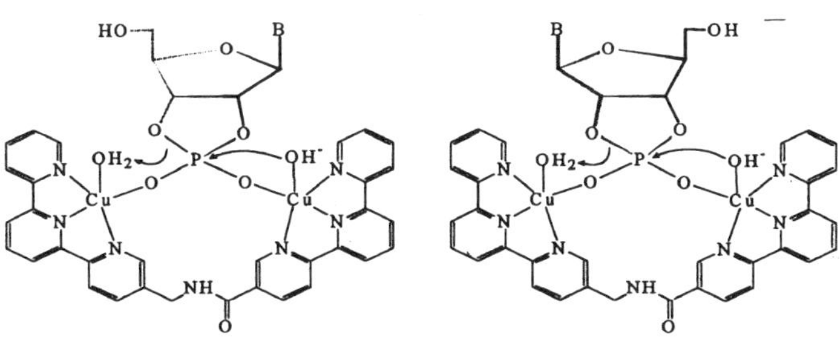

39

appeared to result from hydrolytic cleavage [56]. The authors have reported very fast cleavage of a double-stranded supercoiled plasmide DNA in the presence of the diiron(III) complex $[Fe_2L\,(OH)\,(NO_3)_4]$ of ligand **40** and large excess of H_2O_2 (pH = 8, T = 298 K). Both reactants were required for the cleavage, which produced 5′-phosphate and 3′-OH ends. The authors proposed that a (μ-1,2-peroxo) diiron(III) species interacts with the DNA-phosphate backbone and delivers coordinated peroxide to attack the phosphate diester bond via a nucleophilic mechanism. No further details are provided regarding the mechanism although it may show some similarities to that of a La(III)/H_2O adduct [62] (see later in this section).

Fontecave at al. recently reported the efficiency of a (μ-oxo)diiron(III) complex on the hydrolysis of bis(2,4-dinitrophenyl) phosphate (BDNPP) [57]. The Fe–(μ-O) $(phen)_4\,(H_2O)_2\,(NO_3)_4$ complex (phen = 1,10-phenanthroline) retains its structure when dissolved in water and suffers two deprotonations (pK = 5.00 and 6.85) forming a dihydroxo complex. The crystal structure indicated that both Fe(III) ions are coordinated by four nitrogen ions and one hydroxide ion in addition to the bridging oxo-anion. The kinetic study on BDNPP hydrolysis indicated bell-shaped pH-profile, matching the concentration distribution of the monohydroxo complex. At the optimal pH (pH = 6, T = 323 K, $k = 5*10^{-4}$ s^{-1}, [complex] = 0.08 mM) the complex provides ca. 240-fold rate acceleration compared to the uncatalyzed reaction, and three turnovers were proven. According to the proposed mechanism one of the iron centers in **41** acts as Lewis-acid activator while the other provides the nucleophilic hydroxide ion. The presented model shows several interesting similarities (pH optimum, mechanism) to purple acid phosphatase. Note that the structurally similar [tpa(OH)Fe(μ-O)Fe(H_2O)tpa] $(ClO_4)_3$ complex (tpa = tris(2-pyridylmethyl)amin) has also been reported to hydrolyze triphenyl phosphate to diphenyl phosphate [58].

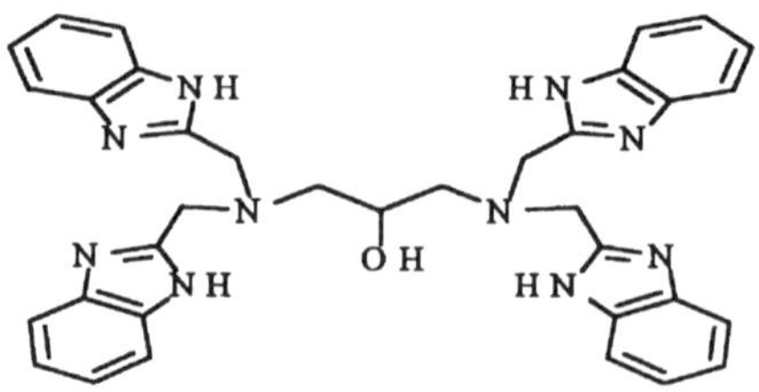

40

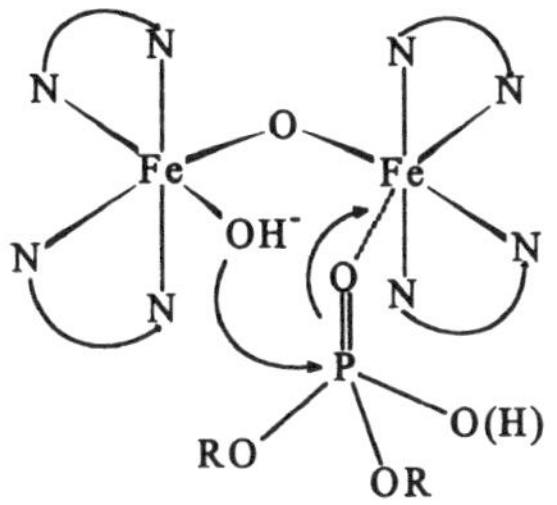

41

Dilanthanide(III) Complexes

Due to the very high efficiency of lanthanide ions on phosphate ester hydrolysis, the number of published papers is exponentially increasing in this field. Most of the authors agree in the fundamental role of oligomeric–hydroxo clusters in the catalytic mechanism. Chin proposes that the $[La_2(OH)_5]^+$ is the active species in the cleavage of ApA: the reported half-life is 13 s at pH 9, 298 K, and 2 mM La(III) concentration [59a]. Several attempts have been made to identify the active species [59] but the results are not consistent. In this way, neither of the proposed mechanisms is generally accepted, although the enormous hydrolytic power of lanthanide ions suggests multifunctional catalysis. The synergetic effect of two different metal ions (lanthanide or non-lanthanide ions) in heterometallic–hydroxo clusters was also demonstrated, but the mechanistic details are even less clear in these cases [60].

More exact picture of catalytic mechanism can be expected when the lanthanide ion is complexed by chelating ligands. In many cases, however, this results in a reduced activity [61]. In case of the early lanthanide(III) ions, substantial acceleration has been reported for the hydrolysis of BNPP [62] and ApA [63] in presence of H_2O_2. $La_2(O_2)_2]_x$ aggregates have been proposed to be the active species for BNPP hydrolysis, and $[La_2(O_2)_3]_x$ for the ApA cleavage. This was explained by the different catalytic requirements of the two substrates. Additional general acid catalysis may be needed for UpU cleavage. Since ^{18}O from labeled H_2O_2 was incorporated into inorganic phosphate formed by cleavage of BNPP, the formation of a peroxophosphate intermediate was proposed which is converted to phosphate with excess H_2O_2. Thus, although hydrolytic products were observed, they probably formed through a nonhydrolytic pathway.

Bruice et al. designed a phosphonate diester with attached chelate ligands for coordination of two metal ions [64]. The bimetallic cooperativity in the La(III) dimer **42** provides extraordinarily high rate enhancement (10^{13}) of the (self) hydrolysis compared to the uncatalyzed reaction. The role of the metal ions in the mechanism is proposed to be (1) to offer double Lewis-acid activation, (2) to provide intramolecular nucleophilic hydroxide ion, (3) to stabilize the transition state, and (4) to facilitate the departure of the leaving alcohol. In spite of the high self-cleaving efficiency, the hydrolysis of the dimer was further promoted by additional La(III) ion in a pH-dependent manner. This is probably due to the weak association of a third La(III) with the phosphonate ester in **42** which also provides a La(III)–OH nucleophile. In this way, the presented data indicate trimetallic cooperativity in the hydrolysis. The hydrolytically inert monomer complex **43** also undergoes hydrolysis in the presence of an additional La(III) ion, providing further details of the mechanism.

Komiyama et al. have described [65] enhanced hydrolysis of ApA by La(III) ion in presence of the tetrapyridine ligand **44**. The complexation was followed by ^{1}H NMR spectroscopy. Besides a 1/1 (La(III)/L) complex, a 2/1 species is formed in the presence of excess metal. Both the 1/1 and 2/1 complexes are efficient promotors of ApA cleavage: the monomer complex being ca. 7, the dimer ca. 70-fold more reactive than the metal ion alone under the same conditions (pH = 7.2, T = 323 K).

Schneider and coworkers have reported on the hydrolysis of BNPP by Pr(III) in the presence of the potentially dinucleating ligand **45** [66]. An aqueous solution of a 2:1 Pr(III)/**45** complex which was prepared in organic solvent is 70 times more reactive toward BNPP than the metal salt alone at pH 7.0 and 323 K. The rate enhancement over spontaneous hydrolysis is 5×10^{6}-fold. The authors suggest cooperation of two metal ions, but there is no direct evidence for the presence of a dinuclear–Pr complex in aqueous solution. Catalytic turnover was not demonstrated.

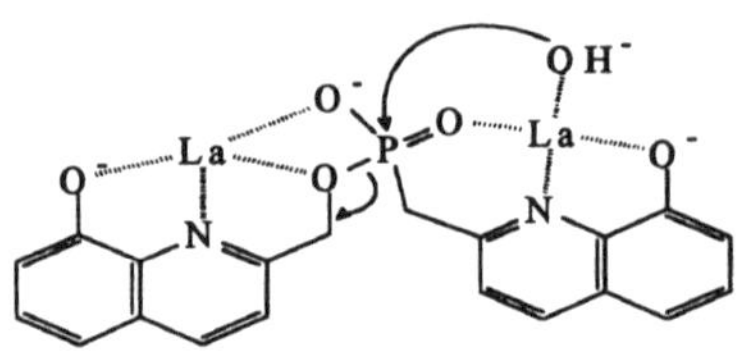

42

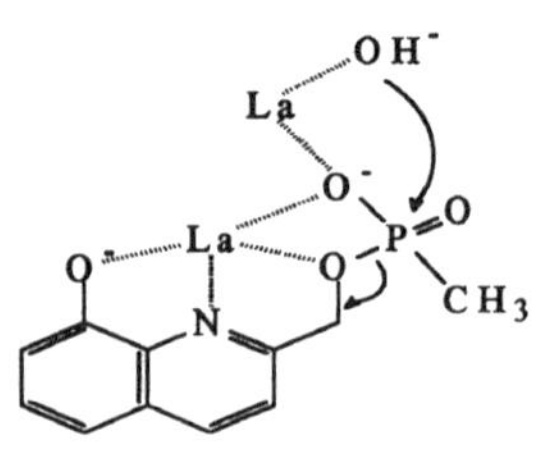

43

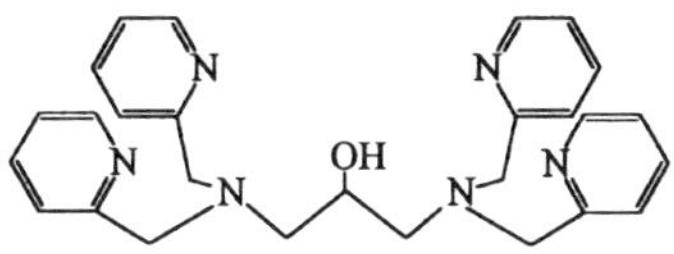

44

4. CONCLUSION

In the last 15 years a number of synthetic models for dinuclear phosphoesterase enzymes have been described which provide solid evidence for the efficient cooperation of two metal ions in the hydrolysis of phosphate esters. A comparison of reactivities of selected dinuclear complexes and their mononuclear analogues is given in Table 2. Some complexes discussed in this article are the most reactive nonenzymatic reagents for phosphoester cleavage and provide rate enhancements up to 12 orders of magnitude under mild conditions, although these systems do not show catalytic turnover. There is a wide variation in the efficiency of metal–metal cooperation, depending strongly on steric preorganization of the metal ions.

Mechanistic features of some low molecular weight model complexes show similarities to the mechanism proposed for certain dinuclear enzymes. In particular, 1,3-bridging of the phosphoester substrate combined with nucleophilic attack of metal-bound hydroxide provides high reactivity both in model compounds and in several phosphomonoesterases. In contrast, a 1,1-bridging mode of the substrate with additional coordination of the leaving group oxygen, as suggested for the 3′,5′-exonuclease subunit of DNA–polymerase I and for other enzymes, has not yet been mimicked by model complexes.

Only for a very few dinuclear phosphoesterase mimics has catalytic turnover been demonstrated. Catalysis may be prevented either by slow kinetics of substrate/product exchange, as in the case of cobalt(III)

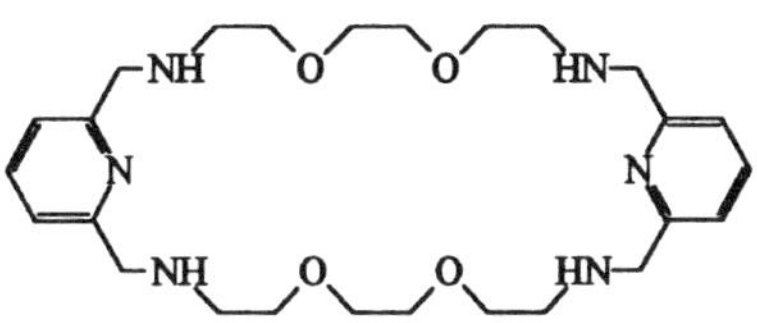

45

Table 2. Selected Dinuclear Phosphoesterase Model Complexes with Rate Increases by Metal–Metal Cooperation over Related Mononuclear Species

Compl.	*Metals*	*Substrate*	*Rate incr. by M2*[a]	*Ref.*
2	$2Co^{3+}$	NPP	26	[31]
6	$2Co^{3+}$	NPP	3	[36]
8	$2Co^{2+}$	BNPP	10	[37]
14	$2Zn^{2+}$	NPP[b]	7	[43]
16	$2Zn^{2+}$	BNPP[b]	5	[43]
19	$2Zn^{2+}$	BNPP	5	[44]
22	$2Zn^{2+}$	UpU[b]	[c]	[45]
23	$3Zn^{2+}$	UpU[b]	[c,d]	[46]
25	$2Zn^{2+}$	HPNP	50	[48]
34	$2Cu^{2+}$	HPNP	26	[52]
36	$2Cu^{2+}$	2′,3′-cAMP	287	[53]
36	$2Cu^{2+}$	ApA	524	[53]
37	$2Cu^{2+}$	HPNP	26	[54]
37	$2\ Cu^{2+}$	HPNP	18	[54]
38	$2Cu^{2+}$	2′,3′-cGMP[b]	457[d]	[55]
38	$2Cu^{2+}$	2′,3′-cAMP[b]	64	[55]
42	$2La^{3+}$	[e]	[c]	[64]
44	$2La^{3+}$	ApA	10	[65]

Notes: [a]k_{obs} (dimer)/k_{obs} (monomer) where "dimer" is the dinuclear complex and monomer is a related monomer complex.
[b]Several substrates were measured.
[c]The effect of monomer complex is too small to be detected.
[d]Important regioselectivity was detected.
[e]Self-hydrolysis.

complexes, or by thermodynamic product inhibition in case of kinetically labile complexes.

Some model complexes in the reaction with diribonucleotides or cyclic ribonucleotides show remarkable substrate and regioselectivity and thus mimic the specifity of enzymes in these respects.

ACKNOWLEDGMENTS

T.G. thanks the Alexander von Humboldt Foundation for a research fellowship. This publication was supported by the Deutsche Forschungsgemeinshaft and by the Hungarian Research Foundation (Project No. T025114).

NOTE ADDED IN PROOF

Several papers related to the present topic have appeared in the literature after the submission of our manuscript, indicative of the great interest in this area. A peroxo-bridged dinuclear cobalt(III) complex, similar to **9**, has been shown to efficiently hydrolyze the nonactivated methyl phenylphosphate [67]. Using ^{18}O-labeled D_2O or O_2, incorporation of ^{18}O to the product was only observed from the solvent, indicating a true hydrolytic mechanism.

Sequence-selective hydrolysis of RNA by a conjugate of a dinuclear Zn^{2+} complex and DNA oligomer have been reported for the first time [68]. The same conjugate with only one equivalent of zinc(II) ion or its mononucleating analogue, as well as the dinuclear complex alone without the DNA tail are virtualy inactive. Notable hydrolytic activity towards UpU has been demonstrated by the synergetic cooperation between Fe^{3+} and Zn^{2+} in an alkoxo-bridged, heterodinuclear complex, providing both a structural and a functional model for several phosphatases [69].

A new imidazole-functionalized calix[4]arene ligand, able to form a dinuclear Cu^{2+} complex, has been reported to hydrolyze HPNP and ethyl *p*-nitrophenylphosphate [70]. The dinuclear complex was found to be 22- and 330-fold more reactive than the corresponding monomer towards the above substrates, respectively. Dinuclear Cu^{2+} complexes of linked triazacyclononane ligands are reported to promote the hydrolysis of the monoribonucleotide GpppG, a model for the 5′-cap structure of mRNA [71]. The dinuclear complexes offer some 100-fold higher reactivity compared to the mononuclear Cu^{2+}-triazacyclononane system.

REFERENCES

[1] Sträter, N., Lipscomb, W.N., Klabunde, T. and Krebs, B., Angew. Chem., 108 (1996) 2158.
[2] Gani, D. and Wilkie, J., Struct. Bonding, 89 (1997) 134.
[3] Wilcox, D.E., Chem. Rev., 96 (1996) 2435.
[4] Kim E.E. and Wyckoff, H.W., J. Mol. Biol., 218 (1991) 449.
[5] Butler-Ransohoff, J.E., Rokita, S.E., Kendall, D.A., Banzon, J.A., Carano, K.S., Kaiser, E. and Matlin, A.R., J. Org. Chem., 57 (1992) 142.
[6] Sträter, N., Klabunde, T., Tucker, P., Witzel, H. and Krebs, B., Science, 268 (1995) 1489.
[7] Klabunde, T., Sträter, N., Witzel, H. and Krebs, B., J. Mol. Biol., 259 (1996) 737.
[8] Goldberg, J.H., Huang, Y., Kwon, P., Greengard, A.C., Nairna, J. and Kuriyan, N., Nature, 376 (1995) 745.

[9] Chu, Y., Lee, E.Y. and Schlender, K.K., J. Biol. Chem., 271 (1996) 2574.
[10] Griffith, J.P., Kim, J.L., Kim, E.E., Sintchak, M.D., Thomson, J.A., Fitzgibbon, M.J., Flemming, M.A., Caron, P.R., Hsiao, K. and Navia, M.A., Cell, 82 (1995) 507.
[11] Bone, R., Frank, L., Springer, J.P. and Atack, J.R., Biochemistry, 33 (1994) 9468.
[12] Zhang, Y., Liang, J., Huang, S., Ke, H. and Lipscomb, W.N., Biochemistry, 32 (1993) 1844.
[13] Freemont, P.S., Friedman, J.M., Beese, L.S., Sanderson, L. and Steitz, T.A., Proc. Natl. Acad. Sci. USA, 85 (1988) 8924–8928.
[14] Beese, L.S. and Steitz, T.A., EMBO J., 10 (1991) 25–33.
[15] Steitz, T.A. and Steitz, J.A., Proc. Natl. Acad. Sci. USA, 90 (1993) 6498–6502.
[16] Hough, E., Hansen, L.K., Birknes, B., Jynge, K., Hansen, S., Hordvık, A., Little, C., Dodson, E. and Derewenda, Z., Nature, 338 (1989) 357–360.
[17] Volbeda, A., Lahm, A., Sakiyama, F. and Suck, D., EMBO J. 10 (1991) 1607–1618.
[18] Benning, M.M., Kuo, J.M., Raushel, F.M. and Holden, H.M., Biochemistry, 34 (1995) 7973.
[19] Guthrie, J.P., J. Am. Chem. Soc., 99 (1977) 3991.
[20] Chin, J., Banaszczyk, B., Jubian, V. and Zou, X., J. Am. Chem. Soc., 111 (1989) 186.
[21] Komiyama, M., J. Biochem., 118 (1995) 665.
[22] Bashkin, J.K., in Karlin, K.D. and Tyeklár, Z. (eds.), *Bioinorganic Chemistry of Copper*, Chapman & Hall, New York, 1993, pp. 132–139.
[23] Meunier, B., *DNA and RATA Cleavers and Chemotherapy of Cancer or Viral Diseases*, Kluwer Academic, 1996.
[24] Häner, R. and Hall, J., Antisense Nucleic Acid Drug Dev., 7 (1997) 423.
[25] Aquino, M.A. , Lim, J. and Sykes, A., J. Chem. Soc. Dalton Trans., (1992) 2135.
[26] Dietrich, M., Münstermann, D., Suerbaum, H. and Witzel, H., Eur. J. Biochem., 199 (1991) 105.
[27] Atack, J.R., Broughton, H.B. and Pollack, S.J., FEBS Lett., 361 (1995) 1.
[28] Pollack, S.J., Atack, J.R., Knowles, M.R., McAllister, G., Ragan, C., Baker, R., Fletcher, S. R., Iversen, L.L. and Broughton, H.B., Proc. Natl. Acad. Sci. USA, 91 (1994) 5766.
[29] Martin, S.F., Wong, Y. and Wagman, A.S., J. Org. Chem., 59 (1994) 4821.
[30] For a critical discussion on the use of nitrophenyl substrates see: Menger, F. M. and Ladika, M., J. Am. Chem. Soc., 109 (1987) 3145; Breslow, R. and Singh, S., Bioorg. Chem. 16 (1988) 408.
[31] Jones, D.R., Lindoy, L.F. and Sargeson, A.M., J. Am. Chem. Soc., 106 (1984) 7807.
[32] Chin, J. and Banaszczyk, M., J. Am. Chem. Soc., 111 (1989) 4103.
[33] Chung, Y., Akkaya, E.U., Venkatachalam, T.K. and Czarnik, A. W., Tetrahedron Lett., 31 (1990) 5413.
[34] Vance, D.H. and Czarnik, A.W., J. Am. Chem. Soc., 115 (1993) 12165.
[35] Seo, J.S., Sung, N.-D., Hynes, R.C. and Chin, J., Inorg. Chem., 35 (1996) 7472.
[36] Wahnon, D., Lebuis, A.-M. and Chin, J., Angew. Chem., 107 (1995) 2594.
[37] Williams, N.H. and Chin, J., J. Chem. Soc., Chem. Commun., (1996) 131.
[38] Morrow, J.R., Buttrey, L.A. and Berback, K.A., Inorg. Chem., 31 (1992) 16.

[39] Kuusela, S. and Lönnberg, H., J. Phys. Org. Chem., 6 (1993) 347.
[40] Jun, J.W., Tanase, T. and Lippard, S.J., Inorg. Chem., 35 (1996) 7590.
[41] Clewley, R.G., Slebocka-Tilk, H. and Brown, R.S., Inorg. Chim. Acta, 157 (1989) 233.
[42] Breslow, R. and Singh, S., Bioorg. Chem., 16 (1988) 408.
[43] Chapman Jr., W.H., and Breslow, R., J. Am. Chem. Soc., 117 (1995) 5462.
[44] Bazzicalupi, C., Bencini, A., Bianchi, A., Fusi, V., Giorgi, C., Paoletti, P., Valtancoli, B. and Zanchi, D., Inorg. Chem., 36 (1997) 2784.
[45] Yashiro, M., Ishikubo, A. and Komiyama, M., J. Chem. Soc., Chem. Comm., (1995) 1793.
[46] Yashiro, M., Ishikubo, A. and Komiyama, M., J. Chem. Soc., Chem. Comm., (1997) 83.
[47] Frey, S.T., Sun, H.H.J., Murthy, N.N. and Karlin, K.D., Inorg. Chim. Acta, 242 (1996) 329.
[48] Molenveld, P., Kapsabelis, J., Engbersen, J.F.J. and Reinhoudt, D. N., J. Am. Chem. Soc., 119 (1997) 2948.
[49] Koike, T., Inoue, M., Kimura, E. and Shiro, M., J. Am. Chem. Soc., 118 (1996) 3091.
[50] Gajda, T., Krämer, R., and Jancsó, A., unpublished results.
[51] Wahnon, D., Hynes, R.C. and Chin, J., J. Chem. Soc., Chem. Comm., (1994) 1441.
[52] Wall, M., Hynes, R.C. and Chin, J., Angew. Chem. Int. Ed. Engl., 32 (1993) 1633.
[53] Young, M.J. and Chin, J., J. Am. Chem. Soc., 117 (1995) 10577.
[54] Liu, S. and Hamilton, A.D., Bioorg. Med. Chem. Lett., 7 (1997) 1785.
[55] Liu, S., Lou, Z. and Hamilton, A.D., Angew. Chem., 109 (1997) 2794.
[56] Schnaith, L.M.T., Hanson, R.S. and Que Jr., L., Proc. Natl. Acad. Sci. USA, 91 (1994) 569.
[57] Dubac-Toia, C., Ménage, S., Vincent, J.-M., Averbuch-Pouchot, A. and Fontacave, M., Inorg. Chem., 36 (1997) 6149.
[58] Hazell, A., Jensen, K.B., McKenzie, C.J. and Toftlund, H., Inorg. Chem., 33 (1994) 3127.
[59] (a) Hurst, P., Takasaki, B.K. and Chin, J., J. Am. Chem. Soc., 118 (1996) 9982; (b) Matsumura, K. and Komiyama, M., J. Biochem., 122 (1997) 387.
[60] See e.g. Takeda, N., Irisawa, M. and Komiyama, M., J. Chem. Soc., Chem. Comm., (1994) 2773; Irisawa, M. and Komiyama, M., J. Biochem., 117 (1995) 465.
[61] See e.g. Schneider, H.-J., Rammo, J. and Hettich, R., Angew. Chem., 105 (1993) 1773.
[62] Takasaki, B. K. and Chin, J., J. Am. Chem. Soc., 117 (1995) 8582.
[63] Kamitani, J., Sumaoka, J., Asanuma, H. and Komiyama, M., J. Chem. Soc. Perkin Trans., 2 (1998) 527.
[64] Tsubouchi, A. and Bruice, T. C., J. Am. Chem. Soc., 117 (1995) 7399.
[65] Yashiro, M., Ishikubo, A. and Komiyama, M., J. Biochem., 120 (1996) 1067.
[66] Ragunathan, K. G., Schneider, H.-J., Angew. Chem., 108 (1996) 1314.
[67] Seo, J.S., Hynes, R.C., Williams, D. and Chin, J., J. Am. Chem. Soc., 120 (1998) 9943.

[68] Matsuda, S., Ishikubo, A., Kuzuya, A., Yashiro, M. and Komiyama, M., Angew. Chem., Int. Ed. Eng., 37 (1998) 3284.
[69] Kamitani, J., Kawahara, R., Yashiro, M. and Komiyama, M., Chem. Lett. (1998) 1047.
[70] Molenveld, P., Engbersen, J.F.J., Kooijman, H., Spek, A.L. and Reinhoudt, D.N., J. Am. Chem. Soc., 120 (1998) 6726.
[71] McCue, K.P., Voss, D.A., Marks, Jr. C. and Morrow, J.R., J. Chem. Soc., Dalton. Trans. (1998) 2961.

SUBJECT INDEX